JAVA BASICS USING CHATGPT / GPT-4

LICENSE, DISCLAIMER OF LIABILITY, AND LIMITED WARRANTY

By purchasing or using this book and companion files (the "Work"), you agree that this license grants permission to use the contents contained herein, including the disc, but does not give you the right of ownership to any of the textual content in the book / disc or ownership to any of the information or products contained in it. *This license does not permit uploading of the Work onto the Internet or on a network (of any kind) without the written consent of the Publisher.* Duplication or dissemination of any text, code, simulations, images, etc. contained herein is limited to and subject to licensing terms for the respective products, and permission must be obtained from the Publisher or the owner of the content, etc., in order to reproduce or network any portion of the textual material (in any media) that is contained in the Work.

MERCURY LEARNING AND INFORMATION ("MLI" or "the Publisher") and anyone involved in the creation, writing, or production of the companion disc, accompanying algorithms, code, or computer programs ("the software"), and any accompanying Web site or software of the Work, cannot and do not warrant the performance or results that might be obtained by using the contents of the Work. The author, developers, and the Publisher have used their best efforts to ensure the accuracy and functionality of the textual material and/or programs contained in this package; we, however, make no warranty of any kind, express or implied, regarding the performance of these contents or programs. The Work is sold "as is" without warranty (except for defective materials used in manufacturing the book or due to faulty workmanship).

The author, developers, and the publisher of any accompanying content, and anyone involved in the composition, production, and manufacturing of this work will not be liable for damages of any kind arising out of the use of (or the inability to use) the algorithms, source code, computer programs, or textual material contained in this publication. This includes, but is not limited to, loss of revenue or profit, or other incidental, physical, or consequential damages arising out of the use of this Work.

The sole remedy in the event of a claim of any kind is expressly limited to replacement of the book and/or disc, and only at the discretion of the Publisher. The use of "implied warranty" and certain "exclusions" vary from state to state, and might not apply to the purchaser of this product.

Companion files for this title are available by writing to the publisher at info@merclearning.com.

Java Basics Using ChatGPT / GPT-4

Oswald Campesato

Mercury Learning and Information
Boston, Massachusetts

Publisher: David Pallai
Mercury Learning and Information
121 High Street, 3rd Floor
Boston, MA 02110
info@merclearning.com
www.merclearning.com
800-232-0223

O. Campesato. *Java Basics Using ChatGPT / GPT-4.*
ISBN: 978-1-501522-437

Library of Congress Control Number: 2023947316

232425321 This book is printed on acid-free paper in the United States of America.

Our titles are available for adoption, license, or bulk purchase by institutions, corporations, etc. For additional information, please contact the Customer Service Dept. at 800-232-0223(toll free).

All of our titles are available in digital format at academiccourseware.com and other digital vendors. Companion files (figures and code listings) for this title are available by contacting info@merclearning.com. The sole obligation of Mercury Learning and Information to the purchaser is to replace the files, based on defective materials or faulty workmanship, but not based on the operation or functionality of the product.

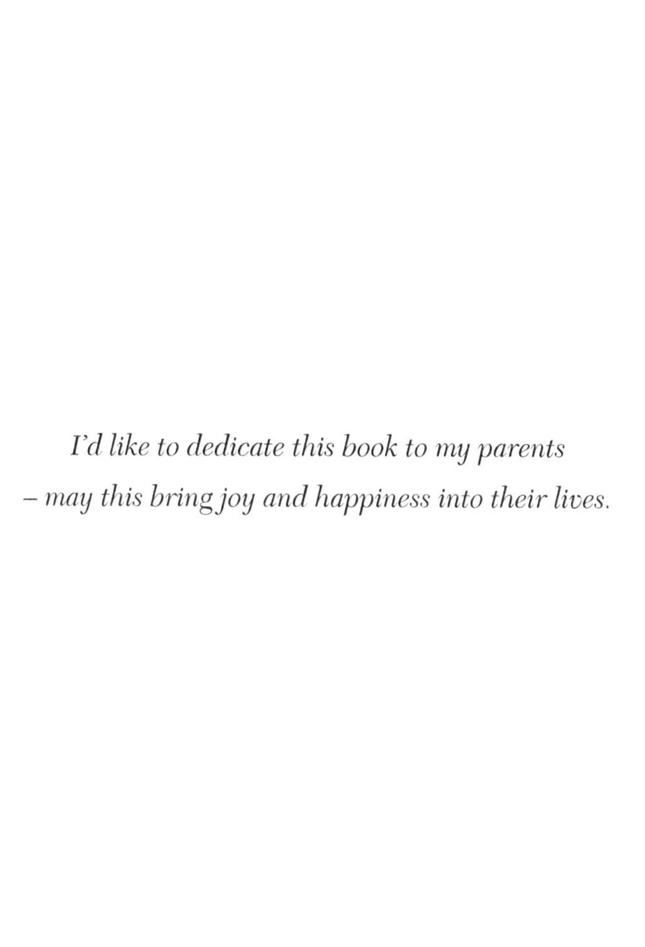

I'd like to dedicate this book to my parents
– may this bring joy and happiness into their lives.

Contents

PREFACE

What Is The Primary Value Proposition For This Book?

This book contains an introduction to `Java` that culminates with a fast-paced view of ChatGPT, during which you can compare the differences and similarities between hand-crafted Java code with ChatGPT-generated Java code. The combination of Java and ChatGPT offers a multi-dimensional approach to problem-solving in programming.

What's The Benefit Of Using ChatGPT With Java?

First and foremost, one of the most compelling reasons for using ChatGPT is the speed at which it can generate code. While hand-written code often requires a substantial amount of time for planning, writing, and debugging, ChatGPT can produce functional code snippets in a matter of seconds. This rapid code generation can be especially useful in the development phase, allowing for quick prototyping and testing of ideas. It's an efficient way to explore multiple solutions to a problem, thereby streamlining the development process.

Secondly, the code generated by ChatGPT serves as an interesting contrast to hand-written code. While hand-written code often reflects the individual styles and best practices adhered to by professional programmers, ChatGPT-generated code offers a different perspective. This can lead to insightful discussions on code efficiency, readability, and maintainability. By comparing and contrasting the two approaches, readers can gain a more well-rounded understanding of programming paradigms and techniques. It's an excellent way to foster a deeper understanding of coding standards and the rationale behind different coding approaches.

Third, ChatGPT's ability to assist in debugging its own generated code adds another layer of utility. If an issue arises in the code produced by the model, ChatGPT can provide suggestions for fixes or even generate corrected code. This self-debugging capability can be a significant time-saver, helping to identify issues more quickly than manual debugging methods. It serves as a valuable educational tool, demonstrating effective debugging techniques and common pitfalls to avoid.

Fourth, ChatGPT can not only generate code but also provide comments and explanations for the code it produces. This is beneficial for educational purposes, as it provides context and understanding, transforming code from mere lines on a page to a comprehensive learning experience.

ChatGPT can generate code in multiple programming languages, offering a polyglot perspective that is invaluable for readers who may not be proficient in a single language. This can make the learning material accessible to a broader audience.

As machine learning models like ChatGPT continue to evolve, their code generation capabilities will only improve, making them increasingly reliable as coding assistants. They can adapt to new languages and frameworks more quickly than traditional methods, making them future-proof to some extent.

In summary, integrating ChatGPT alongside traditional hand-written code samples can offer readers a more dynamic and multifaceted learning experience. The benefits of speed, contrasting viewpoints, debugging assistance, and adaptability make ChatGPT a compelling addition to any technical educational material.

The Target Audience

This book is intended primarily for people who have limited exposure to `Java` and are interested in learning about data structures. This book is also intended to reach an international audience of readers with highly diverse backgrounds in various age groups. While many readers know how to read English, their native spoken language is not English (which could be their second, third, or even fourth language). Consequently, this book uses standard English rather than colloquial expressions that might be confusing to those readers. As such, this book endeavors to provide a comfortable and meaningful learning experience for the intended readers.

What Will I Learn From This Book?

The book covers the basics of Java programming and development environments, including understanding recursion, strings, arrays, fundamental data structures, algorithm analysis, queues and stacks, and follows with the role of ChatGPT in generating, explaining, and debugging code. The book combines hand-crafted Java code with ChatGPT-generated examples for a multifaceted learning experience.

Do I Need To Learn The Theory Portions of This Book?

Once again, the answer depends on the extent to which you plan to become involved in data analytics. For example, if you plan to study machine learning, then you will probably learn how to create and train a model, which is a task that is performed after data cleaning tasks. In general, you will probably need to learn everything that you encounter in this book if you are planning to become a machine learning engineer.

Getting The Most From This Book

Some programmers learn well from prose, others learn well from sample code (and lots of it), which means that there's no single style that can be used for everyone.

Moreover, some programmers want to run the code first, see what it does, and then return to the code to delve into the details (and others use the opposite approach).

Consequently, there are various types of code samples in this book: some are short, some are long, and other code samples "build" from earlier code samples.

What Do I Need To Know For This Book?

Basic knowledge of Java is the most helpful skill, and some exposure to recursion and data structures is obviously helpful. Knowledge of other programming languages can also be helpful because of the exposure to programming concepts and constructs. The less technical knowledge that you have, the more diligence will be required in order to understand the various topics that are covered.

If you want to be sure that you can grasp the material in this book, glance through some of the code samples to get an idea of how much is familiar to you and how much is new for you.

Do The Companion Files Obviate The Need for This Book?

The companion files contain all the code samples to save you time and effort from the error-prone process of manually typing code into a text file. In addition, there are situations in which you might not have easy access to the files. Furthermore, the code samples in the book provide explanations that are not available on the companion files.

All the code samples and figures in this book may be obtained by writing to the publisher at info@merclearning.com.

Does This Book Contain Production-Level Code Samples?

The primary purpose of the code samples in this book is to show you how to use Java in order to solve a variety of programming tasks. Clarity has higher priority than writing more compact code that is more difficult to understand (and possibly more prone to bugs). If you decide to use any of the code in this book in a production environment, you ought to subject that code to the same rigorous analysis as the other parts of your code base.

Another detail to keep in mind is that many code samples in this book contain "commented out" code snippets (often `println()` statements) that were used during the development of the code samples. Those code snippets are intentionally included so that you can uncomment any of those code snippets if you want to see the execution path of the code samples.

What Are The Non-Technical Prerequisites For This Book?

Although the answer to this question is more difficult to quantify, it's very important to have a strong desire to learn about data analytics, along with the motivation and discipline to read and understand the code samples.

How Do I Set Up A Command Shell?

If you are a Mac user, there are three ways to do so. The first method is to use `Finder` to navigate to `Applications > Utilities` and then double click on the `Utilities` application. Next, if you already have a command shell available, you can launch a new command shell by typing the following command:

```
open /Applications/Utilities/Terminal.app
```

A second method for Mac users is to open a new command shell on a MacBook from a command shell that is already visible simply by clicking `command+n` in that command shell, and your Mac will launch another command shell.

If you are a PC user, you can install Cygwin (open source *https://cygwin.com/*) that simulates bash commands, or use another toolkit such as MKS (a commercial product). Please read the online documentation that describes the download and installation process. Note that custom aliases are not automatically set if they are defined in a file other than the main start-up file (such as .bash_login).

What Are The "Next Steps" After Finishing This Book?

The answer to this question varies widely, mainly because the answer depends heavily on your objectives. One possibility involves learning about more complex data structures and implementing them in Java. Another option is to prepare for job interviews involving Java. Yet another possibility is to incorporate ChatGPT (or its successor) in your daily work.

O. Campesato
December 2023

CHAPTER 1

INTRODUCTION TO JAVA

This fast-paced chapter introduces you to an assortment of topics in Java. In many cases explanations are succinct, especially when the purpose of the purpose of the code is intuitive.

The first part of this chapter contains information about downloading Java onto your machine. You will also learn about Java data types and operators, as well as their precedence.

The second section shows you how to create, compile, and launch a Java class from the command line. You will learn how to create a "Hello World" code sample that will be extended in Chapter 2 to help you understand constructors in Java.

The third section discusses numbers, random numbers, and trigonometric functions in Java. The fourth section briefly covers Java characters and strings, and the significance of the new operator. You will also learn how to determine if two strings are equal, and some other useful string-related functions in Java.

A VERY BRIEF INTRODUCTION TO JAVA

Java is an object-oriented programming language that enables you to run your code on many different hardware platforms, including desktops, mobile devices, and the Raspberry PI. You need to install a platform-specific Java distribution for each platform where you intend to launch Java code (see the next section for details).

The following link contains a diagram of the Java 2023 roadmap as well as a layout of Java features:

https://medium.com/javarevisited/the-java-programmer-roadmap-f9db163ef2c2

If you want to learn about the new features that were introduced in different Java releases, navigate to the following website:

https://www.oracle.com/java/technologies/java-se-support-roadmap.html

Downloading a Java Release (Short Version)

In view of the many Java releases that are available, how do you decide which version will be best for you? The answer to this question probably involves one or more of the following:

- the version of Java that your company uses
- the latest LTS (long-term support)
- the version of Java that you prefer
- the latest version of Java
- the features of Java that you need

One other point to keep in mind: you can also download multiple versions of Java onto your machine if you want to address all the points in the preceding bullet list.

After you have decided regarding the version(s) of Java that will work best for you, navigate to the following website and then download the Java distribution for your machine:

https://docs.oracle.com/javase/10/install/toc.htm

SELECTING A VERSION OF JAVA (DETAILED VERSION)

Java 20 is available as this book goes to print, and Java 21 is scheduled for release during September/2023 (and the date is subject to change). Older versions of Java had release cycles that sometimes involves three or more years, whereas newer versions of Java will be released every six months. At the same time, newer releases will have a much more limited set of updates.

Java 8 and Java 11

Java 8 and Java 11 previously had LTS (long term support) status and currently Java 17 has LTS status, as does Java 21. Due to various factors, some larger companies are still working with Java 8, which was one of the most significant releases in the history of Java. If you are interested, the Oracle website contains a list of the features of Java 8. In fact, almost all the features in Java 8 will work correctly in all subsequent releases, up to and including Java 17.

However, Java 8 and Java 11 have been deprecated, and currently Java 17 has LTS status. So, which version of Java is recommended? The answer depends on your requirements. If you are unencumbered with maintaining or developing code for older Java versions, then it's probably safe to work with Java 17. If you don't have a choice, then consider working with Java 13, Java 11, or Java 8 (in this order).

Navigate to the following URL for information regarding earlier versions of Java and release dates for future versions of Java:

https://www.oracle.com/java/technologies/java-se-support-roadmap.html

Java Version Numbers

Up until Java 8, the numbering sequence had the form 1.x. Hence, Java 7 was also called Java 1.7, and Java 8 was called Java 8. However, this older naming convention has been dropped starting from Java 9. For example, if you type the command java -version and you have Java 13 installed on your machine, you will see something similar to the following output:

```
java 13.0.1 2019-10-15
Java(TM) SE Runtime Environment (build 13.0.1+9)
Java HotSpot(TM) 64-Bit Server VM (build 13.0.1+9, mixed
mode, sharing)
```

Aim for a solid grasp of Java 13, and then increase your knowledge by learning about the new features that are available in versions after Java 13.

JRE Versus a JDK

A JRE (Java Runtime Environment) enables you to launch Java code, which means that Java provides the java command-line tool. If you are on a Unix/Linux type of machine, type the following command:

```
$ which java
```

The preceding command displays the following type of output:

```
$ /usr/bin/java
```

On the other hand, a JDK (Java Development Kit) enables you to compile and launch Java code. Although the JRE and JDK were available as separate downloads until Java 8, they have been merged: i.e., starting from Java 9, the Java download is the JDK with the JRE included.

Java Distributions

There are various sites offering the several Java JDK: the OpenJDK project, the OracleJDK, and AdoptOpenJDK. The OpenJDK Project is the only project site that contains the Java source code.

Oracle provides OpenJDK, which is free to use for your Java applications. Updates for older versions are not supported: for example, as soon as Java 13 was available, updates to Java 12 were discontinued. Oracle also provides OracleJDK, which is free to use only during development: if you deploy your code to a production environment, you must pay a fee to Oracle (which will provide you with additional support).

Starting from 2017, you can also use AdoptOpenJDK, which was developed by various developers and vendors, and is freely available (just like OpenJDK).

JAVA IDES

Many people prefer to work in an IDE (integrated development environment), and there are very good IDEs available for many programming languages. For example, if you plan to develop `Android` applications, it's worthwhile to do so in `Android Studio`, which is an alternative for people who are unfamiliar with the command line.

If you want to write `Java` code in an IDE, there are various IDEs that support `Java`, such as NetBeans and Eclipse. However, this book focuses on working with `Java` from the command line, and you can create `Java` classes with a text editor of your choice (or if you prefer, from an IDE). Developing `Java` code from the command line involves simple manual steps, such as updating the `CLASSPATH` environment variable to include `JAR` files and sometimes also including compiled `Java` class files.

This chapter also contains a brief section that shows you how to create a simple `JAR` file that contains compiled `Java` code, and after adding this `JAR` file to the `CLASSPATH` variable, you will see how to invoke your custom `Java` code in your custom `JAR` file. The key point to keep in mind is that the `Java` code samples in this book are short (and often simple) enough that you don't need an IDE.

If you want to write `Java` code in an IDE, there are several IDEs that support `Java`, and the following link contains 10 such IDEs:

https://www.educative.io/blog/best-java-ides-2021

However, the `Java` code samples in this chapter are launched from the command line, and the source code is along with a text editor of your choice for creating the `Java` classes. Working from the command line does involve some additional effort, such as manually updating the `CLASSPATH` environment variable to include `JAR` (`Java` archive) files and compiled `Java` class files.

DATA TYPES, OPERATORS, AND THEIR PRECEDENCE

`Java` supports the usual set of primitive types that are available in many other compiled languages, along with object-based counterparts. For example, the primitive types `int`, `double`, and `float` have class-based counterparts `Integer`, `Double`, and `Float`, respectively.

Some data structures, such as hash tables, do not support primitive data types, which is why class-based counterparts are necessary. For instance, if you need a hash table to maintain a set of integers, you need to create an `Integer`-based object containing each integer and then add that object to the hash table. You then retrieve a specific `Integer`-based value object and extract its underlying `int` value. This approach obviously involves more code than simply storing and retrieving `int` values in a hash table.

Regarding primitive types: Java provides eight primitive data types that are keywords in the language:

- byte: 8-bit signed two's complement integer
- short: 16-bit signed two's complement integer
- int: 32-bit signed two's complement integer
- long: 64-bit signed two's complement integer
- float: single-precision 32-bit IEEE 754 floating point
- Boolean: either true or false
- char: single 16-bit Unicode character

In most cases, the code samples involving arithmetic calculations in this book use integers and floating-point numbers.

Java Comments

Java supports one-line comments as well as multi-line comments, both of which are shown here:

```
// this is a one-line comment
/* this is a multi-line comment
that uses a C-style syntax
// and can include the one-line comment style
*/
```

However, Java does not support nested comments (and neither does C/C++), so this is an error:

```
/* start of first comment /* start of second comment */ end
of first */
```

Based on the earlier examples, the following syntax *is* correct for a comment:

```
/* start of first comment // start of second comment end of
first */
```

Java Operators

Java supports the usual set of arithmetic operators, along with the standard precedence (exponentiation has higher priority than multiplication * and division /, both of which have higher priority than addition + and subtraction -). In addition, Java supports the following operators:

- arithmetic operators: +, -, *, /, %, ++, and --
- assignment operators: =, +=, -=, *=, /=, %=, <<=, >>=, &=, ^=, and !=
- bitwise operators: &, |, ^, and ~
- Boolean operations: `and`, `or`, `not`, and `xor`
- logical operators: &&, ||, and !
- relational operators: ==, !=, >, <, >=, and <=

Java also supports the ternary operator "?" and the operator instanceOf (to check if a variable is an instance of a particular class).

CREATING AND COMPILING JAVA CLASSES

Let's start with the simplest possible Java class that does nothing other than show you the basic structure of a Java class. Listing 1.1 displays the contents of MyClass.java that defines the Java class MyClass.

LISTING 1.1: MyClass.java

```
public class MyClass
{
}
```

The first point to notice is that the name of the file in Listing 1.12 is the same as the Java class. Listing 1.1 defines the Java class MyClass that is qualified by the keywords public and class. You can have more than one class in a Java file but only one of them can be public, and that public class must match the name of the file.

Open a command shell, navigate to the directory that contains MyClass.java, and compile MyClass.java from the command line with the javac executable:

```
javac MyClass.java
```

The javac executable creates a file called MyClass.class that contains Java bytecode. Launch the Java bytecode with the Java executable (do not include any extension):

```
java MyClass
```

Note that you do not specify the suffix .java or .class when you launch the Java bytecode. The output from the preceding command is here:

```
Exception in thread "main" java.lang.NoSuchMethodError:
main
```

NOTE *The name of a file with Java source code must match the public Java class in that file. For example, the file MyClass.java must define a Java class called MyClass.*

Listing 1.2 contains a main method that serves as the initial "entry point" for launching a Java class from the command line, which resolves the error that occurred in Listing 1.1.

LISTING 1.2: MyClass.java

```
public class MyClass
{
   public static void main(String[] args){}
}
```

Listing 1.2 contains the `main()` function whose return type is `void` because it returns nothing. The `main()` function is also `static`, which ensures that it will be available when we launch the `Java` class from the command line. Now compile the code:

```
javac MyClass.java
```

Launch the class from the command line:

```
java MyClass
```

You have now successfully created, compiled, and launched a `Java` class called `MyClass` that still does nothing other than compile successfully. The next step involves displaying a text string in a `Java` class, which is discussed in the next section.

"HELLO WORLD" AND WORKING WITH NUMBERS

Listing 1.3 displays the contents of `HelloWorld1.java` that prints the string "Hello World" from a `Java` class.

LISTING 1.3: HelloWorld1.java

```
public class HelloWorld1
{
   public static void main (String args[])
   {
      System.out.println("Hello World");
   }
}
```

The code in Listing 1.3 is straightforward: the `main()` function contains a single `print` statement that prints the string "`Hello World`". Open a command shell to compile and launch the `Java` code in Listing 1.3:

```
javac HelloWorld1.java
```

Launch the compiled code by typing the following command:

```
java HelloWorld1
```

The output from the preceding command is here:

```
Hello World
```

The companion files contain additional code samples for this chapter that illustrate how to work with numbers, random numbers, trigonometric functions, and bit-wise operators in Java:

- Numbers.java
- RandomNumbers.java
- MathFunctions.java
- TrigFunctions.java
- Bitwise.java

Let's look at some (slightly) more interesting examples of working with numbers in Java.

THE JAVA STRING CLASS

A Java variable of type char is a single character, which you can define via the char keyword like this:

```
char ch1 = 'Z';
```

A Java variable of type String is an *object* that comprises a sequence of char values. As you will see later, Java supports ASCII, UTF8, and Unicode characters. The java.lang.String class implements the Java interfaces Serializable, Comparable, and CharSequence.

The Serializable interface does not contain methods: it's essentially a "marker" interface. The CharSequence interface is for sequences of characters. In addition to the Java String class, the Java StringBuffer class and the Java StringBuilder class implement the CharSequence interface (more details later).

Keep in mind the following point: every Java String is immutable, and in order to modify a Java string (such as append or delete characters), Java creates a new instance "under the hood" for us. As a result, string-related operations are less memory efficient; however, Java provides the StringBuffer and StringBuilder classes for mutable Java strings (also discussed later).

An array of characters is effectively the same as a Java String, as shown here:

```
char[] chars = {'h','e','l','l','o'};
String str1  = new String(chars);
String str2  = "hello";
```

You can access a character in a string with the charAt() method, as shown here:

```
char ch = str1.charAt(1);
```

The preceding code snippet assigns the letter "e" to the character variable ch. You can assign a single quote mark to a variable ch2 with this code snippet:

```
char ch2 = '\'';
```

Listing 1.4 displays the contents of CharsStrings.java that shows you how to define characters, arrays, and strings in Java.

LISTING 1.4: CharsStrings.java

```
public class CharsStrings
{
   public CharsStrings(){}
```

```
    public static void main (String args[])
    {
       char ch1 = 'Z';
       char ch2 = '\'';

       char[] chars = {'h','e','l','l','o'};
       String str1  = new String(chars);
       String str2  = "hello";
       String str3  = new String(chars);
       String str4  = Character.toString(ch2);

       System.out.println("ch1:    "+ch1);
       System.out.println("ch2:    "+ch2);
       System.out.println("chars: "+chars);

       System.out.println("str1:  "+str1);
       System.out.println("str2:  "+str2);
       System.out.println("str3:  "+str3);
       System.out.println("str4:  "+str4);
    }
}
```

Listing 1.4 contains a main() method that initializes the character variable ch2, the character string variable chars, and then the string variables str1 through str4. The remaining portion of the main() method simply prints the values of all these variables. Launch the code in Listing 1.4 and you will see the following output:

```
ch1:   Z
ch2:   '
chars: [C@5451c3a8
str1:  hello
str2:  hello
str3:  hello
str4:  '
```

The output of the `chars` variable is probably different from what you expected. You can use a loop to display the characters in the `chars` variable. Although loops are discussed in Chapter 3, here's a quick preview of a loop that displays the contents of the `chars` variable:

```
System.out.print("chars: ");
for(char ch : chars)
{
   System.out.print(ch);
}
System.out.println();
```

The output from the preceding loop is here:

```
chars: hello
```

Java Strings With Metacharacters

Java treats metacharacters in a string variable in the same manner as any other alphanumeric character. Thus, you don't need to escape metacharacters in Java strings, whereas it's necessary to do so when you define a char variable (see the ch2 variable in the preceding code sample).

You can concatenate two Java strings via the "+" operator, as shown here:

```
String first = "John";
String last  = "Smith";
String full  = first + " " + last;
```

Java treats the following code snippet in a slightly different way:

```
String first = "John";
String last  = "Smith";
first = first + " " + last;
```

The variables first and last are initialized with string values, and then a new block of memory is allocated that is large enough to hold the contents of the variable first, the blank space, and the contents of the variable second. These three quantities are copied into the new block of memory, which is then referenced by the variable full.

Listing 1.5 displays the contents of Strings.java that initializes and then prints several strings.

LISTING 1.5: MyStrings.java

```
public class Strings
{
   public static void main (String args[])
   {
      String str1 = "John", str2 = "Sally";
      String str3 = "*?)",  str4 = "+.@";
      String str5 = "\n\n", str6 = "\t";

      System.out.println("str1: "+str1+" str2: "+str2);
      System.out.println("str3: "+str3+" str4: "+str4);
      System.out.println("str5: "+str5+" str6: "+str6);
   }
}
```

Listing 1.5 defines the Java class MyStrings and a main() function that initializes the strings str1 through str6 with names, metacharacters, and whitespaces. The three println() statements display their values, as shown here:

```
str1: John str2: Sally
str3: *?) str4: +.@
str5:


 str6:
```

There is no visible difference between a space and a tab character in the preceding output. However, redirect the output to a file called out1 (or some other convenient name), open the file in the vi editor, issue the setting ":set list", and you will see the following output:

```
str1: John str2: Sally$
str1: John str2: Sally$
str3: *?) str4: +.@$
str5: $
$
 str6: ^I$
```

The $ symbol indicates the end of a line in the preceding output block. Notice that the last line displays ^I$, which indicates the presence of the tab character that is the value of the str6 variable in Listing 1.5.

THE JAVA new OPERATOR

The previous section contains several examples of directly assigning a string in Java. Another way to do so involves the new operator, which superficially looks the same, but it has an important distinction. When you initialize two variables with the same string, in the manner shown in Listing 1.6, they occupy the same memory location. When you use the new operator, the two variables will occupy different memory locations.

Listing 1.6 displays the contents of ShowPeople.java that uses the new operator to initialize two strings and then print their values.

LISTING 1.6: ShowPeople.java

```
public class ShowPeople
{
   // cannot be initialized in main()
   String str5, str6;

   public ShowPeople(){}

   public static void main (String args[])
   {
      String str1, str2, str3, str4;

      // str1 and str2 occupy different memory locations
      str1 = new String("My name is John Smith");
      str2 = new String("My name is John Smith");

      // str3 and str4 occupy the same memory location
      str3 = "My name is Jane Andrews";
      str4 = "My name is Jane Andrews";

      System.out.println(str1);
```

```
        System.out.println(str2);

        System.out.println(str3);
        System.out.println(str4);


        // error: non-static variable str5 cannot
        // be referenced from a static context:
      //str5 = new String("another string");
      //str6 = new String("yet another string");
    }
}
```

Listing 1.6 contains a `main()` routine that defines and initializes the string variables `str1` through `str4` and then prints their contents. The output from Listing 1.6 is here:

```
My name is John Smith
My name is John Smith
My name is Jane Andrews
My name is Jane Andrews
```

The code in Listing 1.6 is straightforward, but it contains hard-coded values and therefore has no reusability. For example, if you want to print all the names that are defined in arrays containing the names people. Perhaps you want to print names that are randomly selected from those arrays. You might even want to get a person's first name and last name from the command line and then print the person's name. Later in this chapter you will learn how to work with arrays of Strings using various loop constructs, and after that you will learn how to define accessors and mutators for "getting" and "setting" values of variables.

The next section provides more details regarding the difference between "==" and the `equals()` method in the `Java String` class.

EQUALITY OF STRINGS

Unlike other languages, the "==" operator does not determine whether or not two strings are identical: this operator only determines if two variables are referencing the same *memory* location. The `equals()` method will compare the *content* of two strings whereas == operator matches the *object* or reference of the strings.

Listing 1.7 displays the contents of `EqualStrings.java` that illustrates how to compare two strings and determine if they have the same value or the same reference (or both).

LISTING 1.7: EqualStrings.java

```
import java.io.IOException;

public class EqualStrings
```

```
{
    public static void main(String[] args) throws
    IOException
    {
        String str1 = "Pizza";
        String str2 = "Pizza";

        if (str1.equals(str2))
        {
            System.out.println("str1 and str2: equal values");
        }

        if (str1 == str2)
        {
            System.out.println("str1 and str2: equal
            references");
        }

        System.out.println("");

        String str3 = "Pasta";
        String str4 = new String("Pasta");

        if (str3.equals(str4))
        {
            System.out.println("str3 and str4: equal values");
        }
        else
        {
            System.out.println("str3 and str4: unequal
            values");
        }

        if (str3 == str4)
        {
            System.out.println("str3 and str4: equal
            references");
        }
        else
        {
            System.out.println("str3 and str4: unequal
            references");
        }
    }
}
```

Listing 1.7 defines the `Java` class `EqualStrings` and a `main()` function that defines the string variables `str1`, `str2`, `str3`, and `str4`. Launch the code in Listing 1.7 and you will see the following output:

```
str1 and str2: equal values
str1 and str2: equal references
```

```
str3 and str4: equal values
str3 and str4: unequal references
```

Remember: when you create a string literal, the JVM (Java virtual machine) checks for the presence of that string in something called the "string constant pool." If that string exists in the pool, then Java simply returns a reference to the pooled instance; otherwise, a new string instance is created (and it's also placed in the pool).

In order to determine whether or not two strings are identical in Java, use the compareTo(str) method, an example of which is discussed in the next section.

Comparing Strings

Listing 1.8 displays the contents of CompareStrings.java that illustrates how to compare two strings and determine if they have the same value or the same reference (or both).

LISTING 1.8: CompareStrings.java

```
public class CompareStrings
{
   public CompareStrings(){}

   public static void main (String args[])
   {
      String line1 = "This is a simple sentence.";
      String line2 = "this is a simple sentence.";

      System.out.println("line1: "+line1);
      System.out.println("line2: "+line2);
      System.out.println("");

      if (line1.equalsIgnoreCase(line2)) {
         System.out.println(
           "line1 and line2 are case-insensitive equal");
      } else {
         System.out.println(
           "line1 and line2 are case-insensitive
           different");
      }

      if (line1.toLowerCase().equals(line1)) {
         System.out.println("line1 is all lowercase");
      } else {
         System.out.println("line1 is mixed case");
      }
   }
}
```

Listing 1.8 defines the Java class CompareStrings and a main() method that that defines the string variables line1 and line2. Launch the code in Listing 1.8 and you will see the following output:

```
line1: This is a simple sentence.
line2: this is a simple sentence.

line1 and line2 are case-insensitive same
line1 is mixed case
```

SEARCHING FOR A SUBSTRING IN JAVA

Listing 1.9 displays the contents of SearchString.java that illustrates how to use the indexOf() method to determine whether or not a string is a substring of another string. Note that this code sample involves an if statement, which is discussed in more detail in the final code sample of this chapter.

LISTING 1.9: SearchString.java

```
public class SearchString
{
   public SearchString(){}

   public static void main (String args[])
   {
      int index;
      String str1 = "zz", str2 = "Pizza";

      index = str2.indexOf(str1);
      if(index < 0)
      {
         System.out.println(str1+" is not a substring of
         "+str2);
      }
      else
      {
         System.out.println(str1+" is a substring of
         "+str2);
      }
   }
}
```

Listing 1.9 contains a main() method that initializes the string variables str1 and str2, and then initializes the string variable index with the index position of the string str1 in the string str2. If the result is negative then str1 is not a substring of str2, and a message is printed. If str1 *is* a substring of str2, then a corresponding message is displayed. Launch the code in Listing 1.9 and you will see the following output:

```
zz is a substring of Pizza
```

USEFUL STRING METHODS IN JAVA

The Java String class supports a plethora of useful and intuitively-named methods for string-related operations, including: compare(), compareTo(), concat(), equals(), intern(), length(), replace(), split(), and substring().

The Java String class has some very useful methods for managing strings, some of which are listed here (and one of which you have seen already):

- substring(idx1, idx2): the substring from index idx1 to idx2
- compareTo(str): compare a string to a given string
- indexOfStr(str): find the index of a string in another string
- lastIndexOfStr(str): find the index of the last occurrence of a string in another string

Listing 1.10 displays the contents of CapitalizeFirstAll.java that illustrates how to use the substring() method in order to capitalize the first letter of each word in a string, how to convert the string to all lowercase letters, and how to convert the string to all uppercase letters. Note that this example uses a loop, but the code execution in this code sample should be straightforward.

LISTING 1.10: CapitalizeFirstAll.java

```
public class BooleanExamples
{
   public CapitalizeFirstAll(){}

   public static void main (String args[])
   {
      String line1 = "this is a SIMPLE sentence.";
      String[] words = line1.split(" ");
      String line2 = "", first = "";

      for(String word: words)
      {
         first = word.substring(0,1).toUpperCase()+word.
         substring(1);
         line2 = line2 + first + " ";
      }

      System.out.println("line1: "+line1);
      System.out.println("line2: "+line2);

      String upper = line1.toUpperCase();
      String lower = line1.toLowerCase();

      System.out.println("Lower: "+lower);
      System.out.println("Upper: "+upper);
   }
}
```

Listing 1.10 defines the `Java` class `CapitalizeAllFirst` and a `main()` method that initializes the `String` variable `line1` with a text string. The next portion of the `main()` method splits ("tokenizes") the contents of `line1` into an array of words via the `split()` method.

Next, a loop iterates through each word in the `words` array, and sets the variable `first` equal to the uppercase version of the first letter of the current word, concatenated with the remaining letters of the current word, as shown here:

```
first = word.substring(0,1).toUpperCase()+word.
substring(1);
```

The loop also concatenates `first` and a blank space to the string `line2`, which will consist of the words from the variable `line`, with the first letter in each word in uppercase.

The final portion of the `main()` function displays the contents of the modified sentence, along with a lowercase version of the sentence, followed by an uppercase version of the original sentence. Launch the code in Listing 1.16 and you will see the following output:

```
line1: this is a SIMPLE sentence.
line2: This Is A SIMPLE Sentence.
Lower: this is a simple sentence.
Upper: THIS IS A SIMPLE SENTENCE.
```

Parsing Strings in Java

`Java` supports command-line arguments, which enables you to launch programs from the command line with different command line values. Compare `Java` strings with the `equals()` method to determine whether or not the values are the same. Although some languages compare strings via "==", in Java the "==" only compares the two references (not the values) of two strings.

Use the `parseInt()` method to (attempt to) convert a string to an integer:

```
int num = Integer.parseInt("1234");
```

Use the `substring()` method to split a string with white space characters:

```
String[] strArray = aString.split("\\s+");
```

The regular expression `\s+` matches one or more occurrences of various white space characters, including " ", "\t", "\r", and "\n".

The following code snippet reverses a `Java` string via the `StringBuilder` class:

```
String rev = new StringBuilder(original).reverse().
toString();
```

The companion disc contains `OverrideToString.java` that shows you how to override the default toString() method in Java.

Thus far you have seen various examples of working with strings in Java, and for more information, navigate to this URL:

https://docs.oracle.com/javase/7/docs/api/java/lang/String.html

CONDITIONAL LOGIC IN JAVA

Conditional logic enables you to make decisions based on a condition. The simplest type involves a simple "if" statement but can also use if-else as well as if-else-if statements. Another Java construct is the switch statement, which also involves conditional logic, and some tasks are handled more easily with a switch statement instead of multiple if-else statements.

Listing 1.11 displays the contents of `Conditional1.java` that divides two numbers by small integers and uses the integer-valued remainder to print various messages.

LISTING 1.11: Conditional1.java

```
public class Conditional1
{
   int x = 12, y = 15;

   public Conditional1() {}

   public void IfElseLogic()
   {
      if(x % 2 == 0) {
         System.out.println("x is even: "+x);
      } else if(x % 4 == 0) {
         System.out.println("x is divisible by 4: "+x);
      } else {
         System.out.println("x is odd: "+x);
      }

      if(x % 2 == 0) {
         if(x % 4 == 0) {
            System.out.println("x is divisible by 4: "+x);
         }
      }

      if(y % 3 == 0 && y % 5 == 0) {
         System.out.println("y is divisible by 3 and 5:
         "+y);
      } else if(y % 3 == 0) {
         System.out.println("y is divisible only by 3:
         "+y);
      } else if(y % 5 == 0) {
         System.out.println("y is divisible only by 5:
         "+y);
```

```
        } else {
           System.out.println("y is not divisible by 3 or 5:
           "+y);
        }
     }

     public static void main(String args[])
     {
        Conditional1 c1 = new Conditional1();
        c1.IfElseLogic();
     }
}
```

Listing 1.11 defines the Java class `Conditional1` and the public method `IfElseLogic` that performs modulo arithmetic on the integer variables x and y that are initialized with the values of 12 and 15, respectively.

If `x % 2` equals 0, then x is even; if `x % 4` equals 0, then x is a multiple of 4; if `y % 3` equals 0, then y is a multiple of 3. Listing 1.11 contains some if-else code blocks and also a nested block of conditional logic, and you perform a visual calculation in order to compare your results with the generated output.

Launch the code in Listing 1.11 and you will see the following output:

```
x is even: 12
x is divisible by 4: 12
y is divisible by 3 and 5: 15
```

You can create much more complex Boolean expressions involving multiple combinations of "and," "or," and "not" (this involves "!="). Use the code in Listing 1.1 as a baseline and then add your own variations.

The companion files contain BooleanExamples.java that shows you various types of Boolean expressions in Java.

DETERMINING LEAP YEARS

Listing 1.12 displays the contents of `LeapYear.java` that determines whether or not a given year is a leap year. In case you have forgotten, a leap year must be a multiple of 4. However, if a given year is a century that is not a multiple of 400, then that year is not a leap year. Hence, 2000 is a leap year, whereas 1700, 1800, and 1900 are not leap years.

This code sample is a nice example of using nested conditional statements. Try to write your own pseudo code and compare it with the contents of Listing 1.12.

LISTING 1.12: LeapYear.java

```
public class LeapYear
{
   public LeapYear() {}
```

```
    // leap years are multiples of 4 except for
    // centuries that are also multiples of 400
    public void checkYear(int year)
    {
       if(year % 4 == 0)
       {
         if(year % 100 == 0)
         {
             if(year % 400 == 0)
             {
                System.out.println(year + " is a leap
                year");
             }
             else
             {
                System.out.println(year + " is not a leap
                year");
             }
          }
          else
          {
             System.out.println(year + " is a leap year");
          }
       }
       else
       {
          System.out.println(year + " is not a leap year");
       }
    }

    public static void main(String args[])
    {
       LeapYear ly = new LeapYear();

       int[] years = {1234, 1900, 2000, 2020, 3000, 5588};

       for (int year : years)
       {
          ly.checkYear(year);
       }
    }
}
```

Listing 1.12 defines the class LeapYear that contains the method checkYear that determines whether or not a given year is a leap year. If you look at the comment block near the beginning of Listing 1.12, you will see that the method checkYear implements the same conditional logic. Launch the code in Listing 1.12 and you will see the following output:

```
1234 is not a leap year
1900 is not a leap year
```

```
2000 is a leap year
2020 is a leap year
3000 is not a leap year
5588 is a leap year
```

FINDING THE DIVISORS OF A NUMBER

Listing 1.13 contains a `while` loop, conditional logic, and the `%` (modulus) operator in order to find the prime factors of any integer greater than 1.

LISTING 1.13: Divisors.java

```
public class Divisors1
{
   public Divisors1() {}

   public void divisors(int num)
   {
      int div = 2;

      System.out.println("Number: "+num);

      while(num > 1)
      {
         if( num % div == 0)
         {
            System.out.println("divisor: "+div);
            num /= div;
         }
         else
         {
            ++div;
         }
      }
   }

   public static void main(String args[])
   {
      Divisors1 d1 = new Divisors1();
      d1.divisors(12);
   }
}
```

Listing 1.13 defines the Java class Divisors1 that contains the method divisors that determines the divisors of a positive integer num. The divisors method contains a while loop that iterates while num is greater than 1.

During each iteration, if num is evenly divisible by div, then the value of div is displayed, and then num is reduced by dividing it by the value of div. If num is not evenly divisible by div, then div is incremented by 1. The output from Listing 1.8 is here:

```
Number: 12
divisor: 2
divisor: 2
divisor: 3
```

CHECKING FOR PALINDROMES

In case you've forgotten, a palindrome is a sequence of digits or characters that are identical when you read them in both directions (i.e., from left to right and from right to left).

For example, the string BoB is a palindrome, but Bob is not a palindrome. Similarly, the number 12321 is a palindrome but 1232 is not a palindrome.

Listing 1.14 displays the contents of `Palindromes1.java` that checks if a given string or number is a palindrome.

LISTING 1.14: Palindromes1.java

```
public class Palindromes1
{
  public Palindromes1() {}

   public void calculate(String str)
   {
      int result = 0;
      int len = str.length();

      for(int i=0; i<len/2; i++)
      {
         if(str.charAt(i) != str.charAt(len-i-1))
         {
            result = 1;
            break;
         }
      }

      if(result == 0)
      {
         System.out.println(str + ": is a palindrome");
      }
      else
      {
         System.out.println(str + ": is not a palindrome");
      }
   }

   public static void main(String args[])
```

```
    {
        String[] names = {"Dave", "BoB", "radar", "rotor"};
        int[] numbers  = {1234, 767, 1234321, -101};

        Palindromes1 pal1 = new Palindromes1();

        for (String name : names)
        {
           pal1.calculate(name);
        }

        for (int num : numbers)
        {
           pal1.calculate(Integer.toString(num));
        }
    }
}
```

Listing 1.14 defines the Java class Palindromes1 that contains the method calculate to determine whether or not a string is a palindrome. The main() method defines the array names that contains a set of strings and the array numbers that contains a set of numbers.

Next, the main() method instantiates the Palindromes1 class and then invokes the calculate() method with each string in the names array. Similarly, the calculate() method is invoked with each number in the numbers array by converting the numeric values to string by means of the Integer.toString() method. Launch the code in Listing 1.14 and you will see the following output:

```
Dave: is not a palindrome
BoB: is a palindrome
radar: is a palindrome
rotor: is a palindrome
1234: is not a palindrome
767: is a palindrome
1234321: is a palindrome
-101: is not a palindrome
```

The next section in this chapter shows you how to combine Java `while` loops with conditional logic (if-else statements) in Java code.

WORKING WITH ARRAYS OF STRINGS

This section shows you how to print the names of people from the `main()` routine, where the first names and last names are stored in arrays.

Listing 1.15 displays the contents of `PersonArray.java` that contains two private String variables for keeping track of a person's first name and last name, along with methods that enable you to access the first name and last name of any person.

LISTING 1.15: PersonArray.java

```
public class PersonArray
{
   private String[] firstNames = {"Jane","John","Bob"};
   private String[] lastNames  = {"Smith","Jones","Stone"};

   public void displayNames()
   {
      for(int i=0; i<firstNames.length; i++)
      {
         System.out.println("My name is "+
                              firstNames[i]+" "+
                              lastNames[i]);
      }
   }

   public static void main (String args[])
   {
      PersonArray pa = new PersonArray();
      pa.displayNames();
   }
}
```

Listing 1.15 defines the Java class PersonArray that contains the method displayNames whose for loop iterates through the string arrays firstNames and lastNames. The loop variable i is used as an index into these arrays to print a person's first name and last name, as shown here:

```
System.out.println("My name is "+
                   firstNames[i]+" "+
                   lastNames[i]);
```

Launch the code in Listing 1.15 and you will see the following output:

```
My name is John Smith
My name is Jane Andrews
```

WORKING WITH THE STRINGBUILDER CLASS

Recall that the Java String class creates an immutable sequence of characters, which can involve a significant amount of memory for many large strings. On the other hand, the Java `StringBuilder` class is an alternative to the String class that supports a mutable sequence of characters.

The `StringBuilder` class has a number of useful methods, each of which has return type `StringBuilder`, as shown here:

- `capacity()`
- `charAt()`
- `delete()`
- `codePointAt()`

- `codePointBefore()`
- `codePointCount()`
- `deleteCharAt()`
- `ensureCapacity()`
- `getChars()`
- `length()`
- `replace()`
- `reverse()`
- `setCharAt()`
- `setLength()`
- `subSequence()`

Listing 1.16 displays the contents of `PersonRandom.java` that uses a randomly generated index into an array of names in order to print a person's first name and last name.

LISTING 1.16: PersonRandom.java

```
public class PersonRandom
{
   private String[] firstNames = {"Jane","John","Bob"};
   private String[] lastNames = {"Smith","Jones","Stone"};

   public void displayNames()
   {
      String fname, lname;
      int index, loopCount=6, maxRange=20;
      int pCount = firstNames.length;

      for(int i=0; i<loopCount; i++)
      {
         index = (int)(maxRange*Math.random());
         index = index % pCount;

         fname = firstNames[index];
         lname = lastNames[index];

         System.out.println("My name is "+
                                   fname+" "+lname);
      }
   }

   public static void main (String args[])
   {
      PersonRandom pa = new PersonRandom();
      pa.displayNames();
   }
}
```

Listing 1.16 is very similar to the code in Listing 1.15: the important difference is that a name is chosen by means of a randomly generated number, as shown here:

```
index = (int)(maxRange*Math.random());
```

Launch the code in Listing 1.16 and you will see the following output:

```
My name is Jane Smith
My name is John Jones
My name is Jane Smith
My name is Jane Smith
My name is Bob Stone
My name is John Jones
```

Notice that there are six output lines even though there are only three people. This is possible because the randomly generated number is always between 0 and 2, which means that we can generate hundreds of lines of output. Of course, there will be many duplicate output lines because there are only three distinct people.

WORKING WITH THE STRINGBUFFER CLASS

The StringBuilder class is similar to the StringBuffer class: as both of them provide an alternative to String class by making a mutable sequence of characters, but they have different behavior in terms of synchronization.

The `StringBuilder` class is not thread-safe, whereas the StringBuffer class guarantees synchronization. Hence, if you need thread-safe code, use the StringBuffer class; if your code has a single thread, use the `StringBuilder` class because it will be faster in most situations.

Listing 1.17 displays the contents of `PersonRandom.java` that uses a randomly generated index into an array of names in order to print a person's first name and last name.

LISTING 1.17: PersonRandom.java

```
public class PersonRandom
{
   private String[] firstNames = {"Jane","John","Bob"};
   private String[] lastNames = {"Smith","Jones","Stone"};

   public void displayNames()
   {
      String fname, lname;
      int index, loopCount=6, maxRange=20;
      int pCount = firstNames.length;

      for(int i=0; i<loopCount; i++)
```

```
        {
            index = (int)(maxRange*Math.random());
            index = index % pCount;

            fname = firstNames[index];
            lname = lastNames[index];

            System.out.println("My name is "+
                                    fname+" "+lname);
        }
    }

    public static void main (String args[])
    {
        PersonRandom pa = new PersonRandom();
        pa.displayNames();
    }
}
```

Listing 1.17 is very similar to the code in Listing 1.16: the important difference is that a name is chosen by means of a randomly generated number, as shown here:

```
index = (int)(maxRange*Math.random());
```

Launch the code in Listing 1.17 and you will see the following output:

```
My name is Jane Smith
My name is John Jones
My name is Jane Smith
My name is Jane Smith
My name is Bob Stone
My name is John Jones
```

Notice that there are six output lines even though there are only three people. This is possible because the randomly generated number is always between 0 and 2, which means that we can generate hundreds of lines of output. Of course, there will be many duplicate output lines because there are only three distinct people.

STATIC METHODS IN JAVA

Static variables are declared in the same manner as instance variables, but with the `static` keyword. If you declare a method as static, that method can only update static variables; however, static variables can be updated in non-static methods. A key point about static methods is that you can reference a static method without instantiating an object; however, nonstatic methods are only available through an instance of a class. Keep in mind the following points about static methods:

- They can call only other static methods.
- They can only access static data.

- Code changes in static methods affect all class instances
- They do not have a "this" reference.

Listing 1.18 displays the contents of StaticMethod.java that illustrates how to define and then invoke a static method in a Java class.

LISTING 1.18: StaticMethod.java

```
public class StaticMethod
{
   public void display1()
   {
      System.out.println("Inside display1");
   }

   public static void display2()
   {
      System.out.println("Inside display2");
   }

   public static void main (String args[])
   {
      System.out.println("Inside main()");

      // cannot invoke non-static method:
    //StaticMethod.display1();

      // this works correctly:
      StaticMethod.display2();


      // this works correctly:
      StaticMethod sm = new StaticMethod();
      sm.display1();
   }
}
```

Listing 1.18 defines the Java class StaticMethod and a "regular" method display1() as well as a static method display2(). The main() method invokes the display2() method without an instance of the StaticMethod class. Next, the main() method initializes the variable sm as an instance of the StaticMethod class, after which sm invokes the method display1() that is not a static method.

Other Static Types in Java

As you saw in the previous section, a static method in a Java class can be invoked without instantiating an object. In addition, you can define a static Java class containing static methods that can be invoked without instantiating an object.

A *static block* (also called a static initialization block) is a set of instructions that is invoked once when a Java class is loaded into memory. A static block is used for initialization before object construction, as shown here:

```
class MyClass
{
   static int x;
   int y;

   static {
      x = 123;
      System.out.println("static block executed");
   }

   // other methods ...
}
```

The code shown in bold in the preceding code block is executed before (any) constructor is executed.

SUMMARY

This chapter introduced you to some Java features, such as some of its supported data types, operators, and the precedence of Java operators. You also learned how to instantiate objects that are instances of Java classes, and how to perform arithmetic operations in a main() method.

Then you got an overview of Java characters and strings, and the significance of the new operator. In addition, you learned how to determine if two strings are equal, and some other useful string-related functions in Java. Finally, you learned about static methods and static blocks in Java.

CHAPTER 2

RECURSION AND COMBINATORICS

This chapter introduces you to the concept of recursion, along with various `Java` code samples, and then an introduction to concepts in combinatorics, such as combination and permutations of objects.

The first part of this chapter discusses recursion, which is the most significant portion of the chapter. The code samples include finding the sum of an arithmetic series (e.g., the numbers from 1 to n), calculating the sum of a geometric series, calculating factorial values, and calculating Fibonacci numbers. Except for the iterative solution to Fibonacci numbers, these code samples do not involve data structures. However, the examples provide a foundation for working with recursion.

If you are new to recursion or if you tend to struggle with algorithms that involve recursion, here is a suggestion: *find some basic tasks that have iterative solutions and solve those tasks via recursive solutions.* The effort involved in finding recursive solutions will provide some useful practice and assist you in becoming more comfortable with recursion.

The second part of this chapter discusses concepts in combinatorics, such as permutations and combinations. Note that a thorough coverage of combinatorics can fill an entire undergraduate course in mathematics, whereas this chapter contains only some rudimentary concepts.

If you are new to recursion, some of this material might be slightly daunting, but don't be alarmed. Usually, several iterations of reading the material and code samples will lead to a better understanding of recursion.

Alternatively, you can also skip the material in this chapter until you encounter code samples later in this chapter that involve recursion. Chapter 3 contains a code sample for binary search that uses recursion, and another code sample that uses an iterative approach, so you can skip the recursion-based example for now. However, if you plan to learn about trees, then you definitely need to learn about recursion, which is the basis for all the code samples in Chapter 5 that perform various tree-related examples.

WHAT IS RECURSION?

Recursion-based algorithms can provide very elegant solutions to tasks that would be difficult to implement via iterative algorithms. For some tasks, such as calculating factorial values, the recursive solution and the iterative solution have comparable code complexity.

As a simple example, suppose that we want to add the integers from 1 to n (inclusive), and let n = 10 so that we have a concrete example. If we denote S as the partial sum of successively adding consecutive integers, then we have the following:

```
S = 1
S = S + 2
S = S + 3
. . .
S = S + 10
```

If we denote S(n) as the sum of the first n positive integers, then we have the following relationship:

```
S(1) = 1
S(n) = S(n-1) + n for n > 1
```

With the preceding observations in mind, the next section contains code samples for calculating the sum of the first n positive integers using an iterative approach and then with recursion.

ARITHMETIC SERIES

This section shows you how to calculate the sum of a set of positive integers, such as the numbers from 1 to n inclusive. The first algorithm uses an iterative approach and the second algorithm uses recursion.

Before delving into the code samples, there is a simple way to calculate the closed form sum of the integers from 1 to n inclusive, which we will denote as S. Then there are two ways to calculate S, as shown here:

```
S = 1 + 2     + 3     + . . . + (n-1) + n
S = n + (n-1) + (n-2) + . . . + 2     + 1
```

There are n columns, and the sum of the two numbers in each column equals (n+1), and therefore the sum of the right-side of the equals sign is n*(n+1). Since the left-side of the equals sign has the sum 2*S, we have the following result:

```
2*S = n*(n+1)
```

Now divide both sides by 2 and we get the well-known formula for the arithmetic sum of the first n positive integers:

```
S = n*(n+1)/2
```

Incidentally, the preceding formula was derived by a young student who was bored with performing the calculations manually: that student was Karl F. Gauss (in third grade).

Calculating Arithmetic Series (Iterative)

Listing 2.1 displays the contents of the `ArithSum.java` that illustrates how to calculate the sum of the numbers from 1 to `n` inclusive using an iterative approach.

LISTING 2.1: ArithSum.java

```
public class ArithSum
{
   public static int calcsum(int num)
   {
      int sum = 0;
      for(int i=1; i<num+1; i++)
      {
         sum += i;
      }

      return sum;
   }

   public static void main(String[] args)
   {
      int sum = 0, max = 20;

      for(int j=2; j<max+1; j++)
      {
         sum = calcsum(j);
         System.out.println("Sum from 1 to "+j+" = "+sum);
      }
   }
}
```

Listing 2.1 starts with a static method `calcsum()` that contains a loop to iterate through the numbers from 1 to num+1 inclusive, where num is the value of the parameter to this function. During each iteration, the variable sum (which was initialized to 0) is incremented by the value of the loop variable, and the final sum is returned.

The next portion of Listing 2.1 defines the `main()` method that initializes the scalar variables sum to 0 and max to 20, followed by a loop that iterates from 2 to max+1. During each iteration, the `calcsum()` method is invoked with the current value of the loop variable, and the returned value is printed. Launch the code in Listing 2.1 and you will see the following output:

```
sum from 1 to 2 = 3
sum from 1 to 3 = 6
sum from 1 to 4 = 10
sum from 1 to 5 = 15
sum from 1 to 6 = 21
sum from 1 to 7 = 28
```

```
sum from 1 to 8 = 36
sum from 1 to 9 = 45
sum from 1 to 10 = 55
sum from 1 to 11 = 66
sum from 1 to 12 = 78
sum from 1 to 13 = 91
sum from 1 to 14 = 105
sum from 1 to 15 = 120
sum from 1 to 16 = 136
sum from 1 to 17 = 153
sum from 1 to 18 = 171
sum from 1 to 19 = 190
sum from 1 to 20 = 210
```

Calculating Arithmetic Series (Recursive)

Listing 2.2 displays the contents of the `ArithSumRecursive.py` that illustrates how to calculate the sum of the numbers from 1 to `n` inclusive using an iterative approach.

LISTING 2.2: ArithSumRecursive.java

```
public class ArithSumRecursive
{
   public static int calcsum(int num)
   {
      if(num == 0)
      {
        return num;
      }
      else
      {
        return num + calcsum(num-1);
      }
   }

   public static void main(String[] args)
   {
      int sum = 0, max = 20;

      for(int j=2; j<max+1; j++)
      {
          sum = calcsum(j);
          System.out.println("Sum from 1 to "+j+" = "+sum);
      }
   }
}
```

Listing 2.2 starts with a static method `calcsum()` that returns the value num if num equals 0, and otherwise recursively invokes the `calcsum()`

method with the value num-1. The sum of the value of num and the result of this invocation is returned.

The next portion of Listing 2.1 defines the `main()` method that initializes the scalar variables sum to 0 and max to 20, followed by a loop that iterates from 2 to max+1. During each iteration, the `calcsum()` method is invoked with the current value of the loop variable, and the returned value is printed.

Calculating Partial Arithmetic Series

Listing 2.3 displays the contents of the `arith_partial_sum.py` that illustrates how to calculate the sum of the numbers from `m` to `n` inclusive, where `m` and `n` are two positive integers such that `m <= n`, using an iterative approach.

LISTING 2.3: ArithPartialSum.java

```
public class ArithPartialSum
{
   public static int calcPartialSum(int m, int n)
   {
      if(n > m)
      {
         return 0;
      }
      else
      {
         return m*(m+1)/2 - (n-1)*(n)/2;
      }
   }

   public static void main(String[] args)
   {
      int sum = 0, max = 8;

      for(int i=2; i<=max; i++)
      {
         for(int j=i+1; j<=max; j++)
         {
           sum = calcPartialSum(j,i);
           System.out.println("Sum from "+i+" to "+j+" = "+sum);
         }
      }
   }
}
```

Listing 2.3 starts with a static method `calcPartialSum()` that contains conditional logic to return 0 if `n >= m`, and otherwise return the arithmetic expression that is displayed in the "else" clause.

The next portion of Listing 2.3 defines the `main()` method that initializes the scalar variables sum to 0 and max to 8, followed by an outer loop that iterates from 2 to max+1. During each iteration, an inner loop iterates from i+1 to max-1, where I is the loop variable of the outer loop. The inner loop involves the `calcPartialSum()` method with the parameters j and ii (j is the index of the inner loop) and then prints the return value. Launch the code in Listing 2.1 and you will see the following output:

```
Sum from 2 to 3 = 5
Sum from 2 to 4 = 9
Sum from 2 to 5 = 14
Sum from 2 to 6 = 20
Sum from 2 to 7 = 27
Sum from 2 to 8 = 35
Sum from 3 to 4 = 7
Sum from 3 to 5 = 12
Sum from 3 to 6 = 18
Sum from 3 to 7 = 25
Sum from 3 to 8 = 33
Sum from 4 to 5 = 9
Sum from 4 to 6 = 15
Sum from 4 to 7 = 22
Sum from 4 to 8 = 30
Sum from 5 to 6 = 11
Sum from 5 to 7 = 18
Sum from 5 to 8 = 26
Sum from 6 to 7 = 13
Sum from 6 to 8 = 21
Sum from 7 to 8 = 15
```

GEOMETRIC SERIES

This section shows you how to calculate the geometric series of a set of positive integers, such as the numbers from 1 to `n` inclusive. The first algorithm uses an iterative approach and the second algorithm uses recursion.

Before delving into the code samples, there is a simple way to calculate the closed form sum of the geometric series of integers from 1 to `n` inclusive, where `r` is the ratio of consecutive terms in the geometric series. Let S denote the sum, which we can express as follows:

```
S   = 1+ r + r^2 + r^3 + . . . + r^(n-1) + r^n
r*S =    r + r^2 + r^3 + . . . + r^(n-1) + r^n + r^(n+1)
```

Now subtract each term in the second row above from the corresponding term in the first row and we have the following result:

```
S - r*S = 1 - r^(n+1)
```

Now factor `S` from both terms on the left side of the preceding equation and we get the following result:

```
S*(1 - r) = 1 - r^(n+1)
```

Now divide both sides of the preceding equation by the term (1-r) to get the formula for the sum of the geometric series of the first n positive integers:

S = [1 - r^(n+1)]/(1-r)

Notice that if r = 1 then the preceding equation returns an infinite value.

Calculating a Geometric Series (Iterative)

Listing 2.4 displays the contents of the GeometricSum.py that illustrates how to calculate the sum of the numbers from 1 to n inclusive using an iterative approach.

LISTING 2.4: GeometricSum.java

```
public class GeometricSum
{
   public static int geomsum(int num, int ratio)
   {
      int partial = 0;
      int power   = 1;
      int sum     = 0;

      for(int i=1; i<num+1; i++)
      {
          partial += power;
          power *= ratio;
          sum += i;
      }

      return partial;
   }

   public static void main(String[] args)
   {
      int ratio = 2, max = 10, prod = 0;

      for(int j=2; j<max+1; j++)
      {
         prod = geomsum(j, ratio);
         System.out.println("Geometric sum for ratio =
         "+ratio+" from 1 to "+j+" = "+prod);
      }
   }
}
```

Listing 2.4 starts with a static method geomsum() that initializes three integer-valued scalar variables. The next portion of code is a loop that iterates through the numbers from 1 to num+1 inclusive, where num is the value of the parameter to this function. During each iteration, the three scalar variables are updated in order to keep track of the partial geometric sum of numbers. After the loop has completed the value of partial is returned.

The next portion of Listing 2.4 defines the `main()` method that initializes the scalar variables ratio, max, and prod to 2, 10, and 0, respectively. Next, a loop iterates from 2 to max+1. During each iteration, the `geomsum()` method is invoked with the current value of the loop variable j and the value of ratio. In addition, the variable prod is initialized with the return value from `geomsum()` and that value is printed. Now launch the code in Listing 2.4 and you will see the following output:

```
geometric sum for ratio= 2 from 1 to 2 = 3
geometric sum for ratio= 2 from 1 to 3 = 7
geometric sum for ratio= 2 from 1 to 4 = 15
geometric sum for ratio= 2 from 1 to 5 = 31
geometric sum for ratio= 2 from 1 to 6 = 63
geometric sum for ratio= 2 from 1 to 7 = 127
geometric sum for ratio= 2 from 1 to 8 = 255
geometric sum for ratio= 2 from 1 to 9 = 511
geometric sum for ratio= 2 from 1 to 10 = 1023
```

Calculating Geometric Series (Recursive)

Listing 2.5 displays the contents of the `GeomSumRecursive.java` that illustrates how to calculate the sum of the geometric series of the numbers from 1 to `n` inclusive using recursion. Note that the following code sample uses tail recursion.

LISTING 2.5: GeoSumRecursive.java

```
public class GeoSumRecursive
{
   public static int geomsum(int num, int ratio, int term, int sum)
   {
      if(num == 1)
      {
         return sum;
      }
      else
      {
         term *= ratio;;
         sum  += term;;
         return geomsum(num-1,ratio,term,sum);
      }
   }

   public static void main(String[] args)
   {
      int max = 10, ratio = 2, sum = 1, term = 1, prod = 0;
```

```
        for(int j=2; j<max+1; j++)
        {
           prod = geomsum(j, ratio, term, sum);
           System.out.println("Geometric sum for ratio =
           "+ratio+" from 1 to "+j+" = "+prod);
        }
    }
}
```

Listing 2.5 starts with a static method `geomsum()` that contains conditional logic to return 0 if `num` equals 1, and otherwise update the variables `term` and `sum` and then return the result of invoking `geomsum()` with the updated values.

The next portion of Listing 2.5 defines the `main()` method that initializes the scalar variables `max`, `ratio`, `sum`, `term`, and `prod` to 10, 2, 1, 1, and 0, respectively. Next, a loop iterates from 2 to `max+1`, and during each iteration, the `geomsum()` method is invoked with the parameters `j`, `ratio`, `term`, and `sum`. The result of this invocation is assigned to the variable `prod`, which is then printed. Now launch the code in Listing 2.5 and you will see the following output:

FACTORIAL VALUES

This section contains three code samples for calculating factorial values: the first code sample uses a loop and the other two code samples use recursion.

As a reminder, the *factorial* value of a positive integer `n` is the product of all the numbers from 1 to `n` (inclusive). Hence, we have the following values:

```
Factorial(1) = 1*1 = 1
Factorial(2) = 2*1 = 2
Factorial(3) = 3*2*1 = 6
Factorial(4) = 4*3*2*1 = 24
Factorial(5) = 5*4*3*2*1 = 120
Factorial(6) = 6*5*4*3*2*1 = 720
Factorial(7) = 7*6*5*4*3*2*1 = 5040
```

If you look at the preceding list of calculations, you can see some interesting relationships among factorial numbers:

```
Factorial(3) = 3 * Factorial(2)
Factorial(4) = 4 * Factorial(3)
Factorial(5) = 5 * Factorial(4)
Factorial(6) = 6 * Factorial(5)
Factorial(7) = 7 * Factorial(6)
```

Based on the preceding observations, it's reasonably intuitive to infer the following relationship for factorial numbers:

```
Factorial(1) = 1
Factorial(n) = n * Factorial(n-1) for n > 1
```

The next section uses the preceding formula in order to calculate the factorial value of various numbers.

Calculating Factorial Values (Iterative)

Listing 2.6 displays the contents of the `Factorial1.py` that illustrates how to calculate factorial numbers using an iterative approach.

LISTING 2.6: Factorial1.py

```
public class Factorial1
{
   public static int factorial(int num)
   {
      int prod = 1;
      for(int i=1; i<num+1; i++)
      {
          prod *= i;
      }

      return prod;
   }

   public static void main(String[] args)
   {
      int prod= 0, max = 15;

      for(int j=1; j<=max; j++)
      {
          prod = factorial(j);
          System.out.println("Factorial "+j+" = "+prod);
      }
   }
}
```

Listing 2.6 starts with a static method `factorial()` that contains a loop to iterate through the numbers from 1 to n inclusive, where n is the value of the parameter to this function. During each iteration, the variable prod (which was initialized to 1) is multiplied by the value of the loop variable I, and the final product is returned.

The next portion of Listing 2.1 defines the `main()` method that initializes the scalar variables prod to 0 and max to 15, followed by a loop that iterates from 1 to max inclusive. During each iteration, the `factorial()` method is invoked with the current value of the loop variable j, and the returned value is assigned to the variable prod and then printed. Launch the code in Listing 2.6 and you will see the following output:

```
factorial 0 = 1
factorial 1 = 1
factorial 2 = 1
factorial 3 = 2
factorial 4 = 6
factorial 5 = 24
```

```
factorial 6 = 120
factorial 7 = 720
factorial 8 = 5040
factorial 9 = 40320
factorial 10 = 362880
factorial 11 = 3628800
factorial 12 = 39916800
factorial 13 = 479001600
factorial 14 = 6227020800
factorial 15 = 87178291200
factorial 16 = 1307674368000
factorial 17 = 20922789888000
factorial 18 = 355687428096000
factorial 19 = 6402373705728000
```

Calculating Factorial Values (Recursive)

Listing 2.7 displays the contents of the `Factorial2.py` that illustrates how to calculate factorial values using recursion.

LISTING 2.7: Factorial2.py

```
public class Factorial2
{
   public static int factorial(int num)
   {
      if(num <= 1)
      {
         return 1;
      }
      else
      {
         return num * factorial(num-1);
      }
   }

   public static void main(String[] args)
   {
      int prod= 0, max = 15;

      for(int j=1; j<=max; j++)
      {
          prod = factorial(j);
          System.out.println("Factorial "+j+" = "+prod);
      }
   }
}
```

Listing 2.7 starts with a static method `factorial()` that contains conditional logic to return 1 if `num` is at most 1, and otherwise return the product of num and result of invoking `factorial()` with the value num-1.

The next portion of Listing 2.7 defines the `main()` method that initializes the scalar variables prod and max to 0 and 15, respectively. Next, a loop iterates from 1 to max inclusive, and during each iteration, the `factorial()` method is invoked with the loop variable j. The result of this invocation is assigned to the variable prod, which is then printed. Launch the code in Listing 2.7 and you will see the same output as the preceding example.

Calculating Factorial Values (Tail Recursion)

Listing 2.8 displays the contents of the `Factorial3.py` that illustrates how to calculate factorial values using tail recursion.

LISTING 2.8: Factorial3.py

```
public class Factorial3
{
   public static int factorial(int num, int prod)
   {
      if(num <= 1)
      {
         return prod;
      }
      else
      {
         return factorial(num-1, num*prod);
      }
   }

   public static void main(String[] args)
   {
      int prod = 0, max = 15;

      for(int j=1; j<=max; j++)
      {
          prod = factorial(j, 1);
          System.out.println("Factorial "+j+" = "+prod);
      }
   }
}
```

Listing 2.8 starts with a static method `factorial()` that contains conditional logic to return `prod` if `num` is at most 1, and otherwise return the product of num and result of invoking `factorial()` with the value n-1 and n*prod.

The next portion of Listing 2.8 defines the `main()` method that initializes the scalar variables prod and max to 0 and 15, respectively. Next, a loop iterates from 1 to max inclusive, and during each iteration, the `factorial()` method is invoked with the loop variable j and the number 1. The result of this invocation is assigned to the variable prod, which is then printed. Launch the code in Listing 2.8 and you will see the same output as the preceding example.

FIBONACCI NUMBERS

Fibonacci numbers have some real-life counterparts, such as the pattern of sunflower seeds and the locations where tree branches form. Here is the definition of the Fibonacci sequence:

```
Fib(0) = 0
Fib(1) = 1
Fib(n) = Fib(n-1)+Fib(n-2) for n >= 2
```

Note that it's possible to specify different "seed" values for `Fib(0)` and `Fib(1)`, but the values 0 and 1 are the most commonly used values.

Calculating Fibonacci Numbers (Recursive)

Listing 2.9 displays the contents of the `Fibonacci1.py` that illustrates how to calculate Fibonacci numbers using recursion.

LISTING 2.9: Fibonacci1.py

```
public class Fibonacci1
{
   // very inefficient:
   public static int fibonacci(int num)
   {
      if(num <= 1)
      {
         return num;
      }
      else
      {
         return fibonacci(num-2) + fibonacci(num-1);
      }
   }

   public static void main(String[] args)
   {
      int fib = 0, max = 20;

      for(int j=0; j<=max; j++)
      {
          fib = fibonacci(j);
          System.out.println("Fibonacci of "+j+" = "+fib);
      }
   }
}
```

Listing 2.9 starts with a static method `fibonacci()` that contains conditional logic to return `num` if `num` is at most 1, and otherwise return the sum of fibonacci(num-2) and fibonacci(n-1).

The next portion of Listing 2.9 defines the `main()` method that initializes the scalar variables fib and max to 0 and 20, respectively. Next, a loop iterates from 1 to max inclusive, and during each iteration, the `fibonacci()` method is invoked with the loop variable j. The result of this invocation is assigned to the variable fib, which is then printed. Launch the code in Listing 2.8 and you will see the following output:

```
fibonacci 0 = 0
fibonacci 1 = 1
fibonacci 2 = 1
fibonacci 3 = 2
fibonacci 4 = 3
fibonacci 5 = 5
fibonacci 6 = 8
fibonacci 7 = 13
fibonacci 8 = 21
fibonacci 9 = 34
fibonacci 10 = 55
fibonacci 11 = 89
fibonacci 12 = 144
fibonacci 13 = 233
fibonacci 14 = 377
fibonacci 15 = 610
fibonacci 16 = 987
fibonacci 17 = 1597
fibonacci 18 = 2584
fibonacci 19 = 4181
```

Calculating Fibonacci Numbers (Iterative)

Listing 2.10 displays the contents of the `Fibonacci2.py` that illustrates how to calculate Fibonacci numbers using an iterative approach.

LISTING 2.10: Fibonacci2.py

```
public class Fibonacci2
{
   public static void main(String[] args)
   {
      int max = 21;
      int arr1[] = new int[max];
      arr1[0] = 0;
      arr1[1] = 1;

      for(int j=2; j<max; j++)
      {
          arr1[j] = arr1[j-1] + arr1[j-2];
          System.out.println("Fibonacci of "+j+" = "+arr1[j]);
      }
   }
}
```

Listing 2.10 defines the `main()` method that initializes the scalar variable max and the integer array arr1 that is initialized with max number of entries. Notice that the first two entries are assigned 0 and 1: you can also use different initial values. Next, a loop iterates from 2 to max, and during each iteration, the `jth` element of arr1 is initialized with the sum of arr1[j-1] and arr1[j-2]. The next code snippet prints the contents of arr1[j]. As you can see, no recursion is used in this code sample. Launch the code in Listing 2.10 and you will see the same output as Listing 2.x in the previous example.

TASK: REVERSE A STRING VIA RECURSION

Listing 2.11 displays the contents of `Reverse.java` that illustrates how to use recursion in order to reverse a string.

LISTING 2.11: Reverse.java

```
public class Reverse
{
   public static String reverser(String str)
   {
      int strLen = str.length();
      if(strLen == 1) {
         return str;
      }

      String lastPart = str.substring(strLen-1,strLen);
      String firstPart = str.substring(0,strLen-1);
     //System.out.println("first: "+firstPart+" last: "+lastPart);
      return lastPart+reverser(firstPart);
   }

   public static void main(String[] args)
   {
      String result = "";
      String[] names = new String[]{"Dave", "Nancy",
      "Dominic"};

      for(int idx=0; idx<names.length; idx++)
      {
        result = reverser(names[idx]);
        System.out.println(
            "=> Word: "+names[idx]+" reverse: "+result+"\n");
      }
   }
}
```

Listing 2.11 starts with a static method `reverser()` that contains conditional logic to return the variable str if `str` is empty or has length 0. Otherwise, a new string called *newts* is constructed that consists of appending the first character in str to the the "all but first" contents of str. The contents of this new string are printed and then the reverser() method is recursively invoked with str.

The next portion of Listing 2.11 defines the `main()` method that initializes the scalar variables result to an empty string and the string-based array names to a set of three string values.

Next, a loop iterates through each element of the names array and invokes the reverser() method with the current string. The result of the method invocation is assigned to the variable result, which is then printed. Launch the code in Listing 2.8 and you will see the following output:

```
all-but-first chars: ['a', 'n', 'c', 'y']
all-but-first chars: ['n', 'c', 'y']
all-but-first chars: ['c', 'y']
all-but-first chars: ['y']
all-but-first chars: []
=> Word:  Nancy  reverse:  ['y', 'c', 'n', 'a', 'N']

all-but-first chars: ['a', 'v', 'e']
all-but-first chars: ['v', 'e']
all-but-first chars: ['e']
all-but-first chars: []
=> Word:  Dave  reverse:  ['e', 'v', 'a', 'D']

all-but-first chars: ['o', 'm', 'i', 'n', 'i', 'c']
all-but-first chars: ['m', 'i', 'n', 'i', 'c']
all-but-first chars: ['i', 'n', 'i', 'c']
all-but-first chars: ['n', 'i', 'c']
all-but-first chars: ['i', 'c']
all-but-first chars: ['c']
all-but-first chars: []
=> Word:  Dominic  reverse:  ['c', 'i', 'n', 'i', 'm', 'o', 'D']
```

TASK: CHECK FOR BALANCED PARENTHESES

This task can arise as a computer science interview question, and here are some examples:

```
S1 = "()()()"
S2 = "(()()())"
S3 = "()("
S4 = "(())"
S5 = "()()("
```

As you can see, S1, S3, and S4 have balanced parentheses, whereas S2 and S5 have unbalanced parentheses.

Listing 2.12 displays the contents of `BalancedParentheses.java` that illustrates how to determine whether or not a string consists of balanced parentheses.

LISTING 2.12: BalancedParentheses.java

```
// https://stackoverflow.com/questions/31849977/balanced-
parenthesis-how-to-count-them
public class BalancedParens
{
   public BalancedParens(){}

   public static void main(String[] args)
   {
      String[] exprs = new String[]{"(){}[]","(()","(())"};

      for(int idx=0; idx<exprs.length; idx++)
      {
         String str = exprs[idx];
         int counter = 0;

         for (char ch : str.toCharArray())
         {
           if (ch == '(') counter++;
           else if (ch == ')') counter--;

           if (counter < 0) break;
         }

         if (counter == 0) {
           System.out.println("balanced string:"+str+"\n");
         } else {
          System.out.println("unbalanced string:"+str+"\n");
         }
      }
   }
}
```

The intuition for the code involves an integer variable counter in order to count parentheses: increment counter for each opening parentheses and decrement counter for each closing parenthesis.

Listing 2.12 defines the `main()` method that initializes the string-based array exprs with three string values consisting of combinations of parentheses and curly braces. Next, an outer loop iterates through each element of the exprs array and then an inner loop iterates through the characters of the current string via the variable ch. The loop contains conditional logic that increments the variable counter if ch equals "(" and decrements the variable counter if ch equals ")". In addition, there is an early exit from the loop if the counter is negative.

The final code portion of Listing 2.12 involves conditional logic: if the counter equals 0, then the current string is well-balanced, otherwise the current string is unbalanced. Launch the code in Listing 2.8 and you will see the following output:

```
balanced string:(){}[]

unbalanced string:((}

balanced string:(())
```

TASK: CALCULATE THE NUMBER OF DIGITS

Listing 2.13 displays the contents of `CountDigits.java` that illustrates how to calculate the number of digits in positive integers.

LISTING 2.13: CountDigits.java

```
public class CountDigits
{
   public CountDigits(){}

   public static int countDigits(int num, int result)
   {
      if( num == 0 )
      {
        return result;
      }
      else
      {
        //print("new result:",result+1)
        //print("new number:",int(num/10))
        return countDigits((int)Math.floor(num/10), result+1);
      }
   }

   public static void main(String[] args)
   {
      int[] numbers = new int[]{1234, 767, 1234321, 101};

      for (int num : numbers)
      {
         int result = countDigits(num, 0);
         System.out.println("Digits in "+num+" = "+result);
      }
   }
}
```

Listing 2.13 starts with a static method `countDigits()` that contains conditional logic to return num if `num` equals 0, and otherwise return the recursive invocation of countDigits() with the integer portion of num/10 and result+1, which are based on the parameters num and result of this recursive function.

The next portion of Listing 2.9 defines the `main()` method that initializes the integer-valued array numbers with four integer values. Next, a loop iterates through the elements of the numbers array, and then invokes the method countDigits() with the current element num and the number 0. The result of the invocation is assigned to the integer-valued variable result, and then the value of result is printed. Launch the code in Listing 2.8 and you will see the following output:

Note that the code in Listing 2.13 can be made even more concise: however, the current listing is slightly easier to understand. Now launch the code in Listing 2.13 and you will see the following output:

```
Digits in  1234  =  4
Digits in  767  =  3
Digits in  1234321  =  7
Digits in  101  =  3
```

TASK: DETERMINE IF A POSITIVE INTEGER IS PRIME

Listing 2.14 displays the contents of `CheckPrime.java` that illustrates how to calculate the number of digits in positive integers.

LISTING 2.14: CheckPrime.java

```
public class CheckPrime
{
   public CheckPrime(){}

   public static int isPrime(int num)
   {
      int PRIME = 1, COMPOSITE = 0;
      int div = 2, num2 = (int)(num/2);

      while(div <= num2)
      {
        if( num % div != 0) {
          div += 1;
        } else {
          return COMPOSITE;
        }
      }

     //print("found prime:",num)
     return PRIME;
   }
```

```
    public static void main(String[] args)
    {
      String result = "";
      int upperBound = 20;
      int[] numbers = new int[]{1234, 767, 1234321, 101};

      for(int num = 2; num < upperBound; num++)
      {
         result = findPrimeDivisors(num);
        System.out.println("Prime divisors of "+num+": "+result);
      }
    }
}
```

Listing 2.14 starts with a static method `isPrime()` with an integer-valued parameter num. The scalar-valued variables PRIME, COMPOSITE, div, and num2 are initialized with the values 1, 0, 2, and the integer portion of num/2, respectively.

Next, a while loop executes as long as the value of div is less than the value of num2. A conditional block of code inside the loop checks of the remainder of num2 divided by div is nonzero: if so, then the variable div is incremented, otherwise the value of COMPOSITE is returned.

The next portion of Listing 2.9 defines the `main()` method that initializes the variables result, upperBound, and numbers to an empty string, the integer 20, and an array of four integer values, respectively. Next, a loop iterates from 2 to upperBound, and during each iteration, the method findPrimeDivisors() is invoked with the value of num. The result of the invocation is assigned to the variable result, which is then printed. Launch the code in Listing 2.8 and you will see the following output:

```
2 : is prime
3 : is prime
4 : is not prime
5 : is prime
6 : is not prime
7 : is prime
8 : is not prime
9 : is not prime
10 : is not prime
11 : is prime
12 : is not prime
13 : is prime
14 : is not prime
15 : is not prime
16 : is not prime
17 : is prime
18 : is not prime
19 : is prime
```

TASK: FIND THE PRIME DIVISORS OF A POSITIVE INTEGER

Listing 2.15 displays the contents of `CheckPrime.java` that illustrates how to find the prime divisors of a positive integer.

LISTING 2.15: CheckPrime.java

```
public class CheckPrime
{
   public CheckPrime(){}

   public static int isPrime(int num)
   {
      int PRIME = 1, COMPOSITE = 0;

      int div = 2, num2 = (int)(num/2);

      while(div < num2)
      {
        if( num2 % div != 0) {
          div += 1;
        } else {
          return COMPOSITE;
        }
      }

     //print("found prime:",num)
     return PRIME;
   }

   public static String findPrimeDivisors(int num)
   {
     int div = 2;
     String prime_divisors = "";

     while(div <= num)
     {
       int prime = isPrime(div);
       if(prime == 1)
       {
         //print("=> prime number:",div)
         if(num % div == 0)
         {
           prime_divisors += " "+div;
           num = (int)(num/div);
           //System.out.println("2prime_divisors:"+prime_
           divisors);
         } else {
           div += 1;
         }
```

```
        } else {
          div += 1;
        }
      }

      return prime_divisors;
   }

   public static void main(String[] args)
   {
      int result = 0, upperBound = 20;
      for(int num = 2; num < upperBound; num++)
      {
         result = isPrime(num);
         if(result == 1) System.out.println(num+" is prime");
         if(result == 0) System.out.println(num+" is not
         prime");
      }
   }
```

Listing 2.15 starts with a static method `isPrime()` that is identical to the method in Listing 2.14.

The next portion of Listing 2.15 defines the `findPrimeDivisors()` method that initializes the variables `div` and `prime_divisors` with the values 2 and "", respectively. The next portion of the code is a `while` loop that executes as long as `div` is less than or equal to `num`, where the latter is the parameter of this method.

During each iteration, the variable `prime` is initialized with the result of invoking the `isPrime()` method with the variable `div`. If `prime` equals 1, then we check if `div` is a divisor of `num`. Specifically, if `num` divided by `div` has remainder 0, we have found a divisor, which we append to the string variable `prime_divisors`, and then set `num` equal to the integer portion of dividing `num` by `div`. However, if `num` divided by `div` is *not* 0, simply increment the value of `div`. In addition, if `prime` does not equal 1, then again, we increment the value of `div`. After the loop has completed execution, return the variable `prime_divisors`, which consists of a string of the prime divisors of the integer `num`.

The next portion of Listing 2.9 defines the `main()` method that initializes the variables `result` and `upperBound` with the values 0 and 20, respectively. Next, a loop iterates from 2 to `upperBound`, and during each iteration, the variable `result` is initialized with the value returned from the method `isPrime()` with the loop variable `num`. If the value of `result` equals 1, then a message is displayed that `num` is a prime number; otherwise, a message is displayed that `num` is *not* a prime number. Launch the code in Listing 2.8 and you will see the following output:

```
Prime divisors of  2 :  2
Prime divisors of  3 :  3
Prime divisors of  4 :  2 2
Prime divisors of  5 :  5
Prime divisors of  6 :  2 3
Prime divisors of  7 :  7
Prime divisors of  8 :  2 2 2
Prime divisors of  9 :  3 3
Prime divisors of  10 :  2 5
Prime divisors of  11 :  11
Prime divisors of  12 :  2 2 3
Prime divisors of  13 :
Prime divisors of  14 :  2 7
Prime divisors of  15 :  3 5
Prime divisors of  16 :  2 2 2 2
Prime divisors of  17 :
Prime divisors of  18 :  2 3 3
Prime divisors of  19 :
```

TASK: GOLDBACH'S CONJECTURE

Goldbach's conjecture states that every even number greater than 4 can be expressed as the sum of two odd prime numbers.

Listing 2.16 displays the contents of `GoldbachConjecture.java` that illustrate how to determine a pair of prime numbers whose sum equals a given even number.

LISTING 2.16: GoldbachConjecture.java

```
public class GoldbachConjecture
{
   public GoldbachConjecture(){}

   public static int isPrime(int num)
   {
      int PRIME = 1, COMPOSITE = 0;
      int div = 2;

      while(div < num)
      {
         if( num % div != 0) {
            div += 1;
         } else {
            return COMPOSITE;
         }
      }
      return PRIME;
   }
```

```
    public static void findPrimeFactors(int even_num)
    {
       int halfway = (int)(even_num/2);
       for(int num=0; num<halfway; num++)
       {
         if(isPrime(num) == 1) {
           if(isPrime(even_num-num) == 1) {
             System.out.println(even_num+" = "+num+" +
             "+(even_num-num));
           }
         }
       }
    }

    public static void main(String[] args)
    {
       int upperBound = 20;
       for(int num=0; num<upperBound; num++) {
         findPrimeFactors(num);
       }
    }
}
```

Listing 2.16 starts with a static method `isPrime()` it that is identical to the method in Listing 2.14.

The next portion of Listing 2.16 defines the `findPrimeFactors()` method that initializes the variable `half_way` as the integer portion of `even_num/2`, where `even_num` is a parameter of this method. The next portion of the code is a loop that executes from 0 to `halfway` with the loop variable `num`. During each iteration, if the result of invoking `isPrime(num)` equals 1, `isPrime(even_num-num)` is checked to see if it also equals 1: if the latter is true, then `even_num` and `num` satisfy Goldbach's conjecture, so their values are printed.

The next portion of Listing 2.9 defines the `main()` method that initializes the variable `upperBound` with the value 20. Next, a loop iterates from 0 to `upperBound`, and during each iteration, the method `findPrimeFactors()` is involved with the current value of the loop variable `num`. Now launch the code in Listing 2.8 and you will see the following output:

```
8  =  3 + 5
10  =  3 + 7
12  =  5 + 7
14  =  3 + 11
16  =  3 + 13
16  =  5 + 11
18  =  5 + 13
18  =  7 + 11
20  =  3 + 17
20  =  7 + 13
22  =  3 + 19
```

```
22  =  5 + 17
24  =  5 + 19
24  =  7 + 17
24  =  11 + 13
26  =  3 + 23
26  =  7 + 19
28  =  5 + 23
28  =  11 + 17
```

TASK: CALCULATE THE GCD (GREATEST COMMON DIVISOR)

Listing 2.17 displays the contents of `GCD.java` that illustrates how to calculate the GCD of two positive integers.

LISTING 2.17: GCD.java

```
public class GCD
{
   public GCD() {}

   public int gcd(int num1, int num2)
   {
      if(num1 % num2 == 0) {
         return num2;
      }
      else if (num1 < num2) {
         System.out.println("Switching "+num1+" and "+num2);
         return gcd(num2, num1);
      }
      else {
         System.out.println("Reducing "+num1+" and "+num2);
         return gcd(num1-num2, num2);
      }
   }

   public static void main(String args[])
   {
      GCD g = new GCD();
      int result = g.gcd(24,10);
      System.out.println("The GCD of 24 and 10 = "+result);
   }
}
```

Listing 2.18 starts with a method `gcd()` that takes two integer-valued parameters `num1` and `num2`. The first section of conditional logic checks if the remainder of `num1` divided by `num2` is zero: if so, then `num2` is returned.

The second section of conditional logic checks if `num1` is less than `num2`: if so, then the method `gcd()` is recursively invoked with the parameter values `num2` and `num1`, and the result is returned.

The third section of conditional logic returns the result of a recursive invocation of the method `gcd()` with the parameter values `num1-num2` and `num2`, and the result is returned.

The next portion of Listing 2.18 defines the `main()` method that instantiates the variable `g` as an instance of the GCD class. Next, the variable `result` is initialized with the result of invoking the `gcd()` method with the value 24 and 10, and then its value is printed. Launch the code in Listing 2.18 and you will see the following output:

```
gcd of 10 and 24 = 2
gcd of 24 and 10 = 2
gcd of 50 and 15 = 5
gcd of 17 and 17 = 17
gcd of 100 and 1250 = 50
```

Now that we can calculate the GCD of two positive integers, we can use this code to easily calculate the LCM (lowest common multiple) of two positive integers, as discussed in the next section.

TASK: CALCULATE THE LCM (LOWEST COMMON MULTIPLE)

Listing 2.18 displays the contents of `LCM.java` that illustrates how to calculate the LCM of two positive integers. Keep in mind that `LCM(x,y) = (x*y)/GCD(x,y)` for any positive integers `x` and `y`.

LISTING 2.18: LCM.java

```
public class LCM
{
   public LCM() {}

   public int gcd(int num1, int num2)
   {
      if(num1 % num2 == 0) {
         return num2;
      }
      else if (num1 < num2) {
         return gcd(num2, num1);
      }
      else {
         return gcd(num1-num2, num2);
      }
   }

   public int lcm(int num1, int num2)
   {
      int gcd1 = gcd(num1, num2);
```

```
        int lcm1 = num1*num2/gcd1;

        return lcm1;
    }

    public static void main(String args[])
    {
        LCM lcm1 = new LCM();
        int result = lcm1.lcm(24,10);
        System.out.println("The LCM of 24 and 10 = "+result);
    }
}
```

Listing 2.18 starts with a method `gcd()` that is identical to the method that is defined in Listing 2.17. The next portion of Listing 2.18 defines the method lcm() that takes parameters num1 and num2. This method the GCD of `num1` and `num2` and assigns the result to the variable `gcd1`. Next, the quantity `num1*num2/gcd1` is calculated and assigned to the variable `lcm1`, which is then returned.

The next portion of Listing 2.18 defines the `main()` method that instantiates the variable `lcm1` as an instance of the LCM class. Next, the variable result is initialized with the result of invoking the `lcm()` method with the value 24 and 10, and then its value is printed. Now launch the code in Listing 2.19 and you will see the following output:

```
The LCM of 24 and 10 = 120
```

This concludes the portion of the chapter regarding recursion. The next section introduces you to combinatorics (a well-known branch of Mathematics),

WHAT IS COMBINATORICS?

In simple terms, combinatorics involves finding formulas for counting the number of objects in a set. For example, how many different ways can five books can be ordered (i.e., displayed) on a bookshelf? The answer involves permutations, which in turn is a factorial value; in this case, the answer is 5! = 120.

As a second example, suppose how many different ways can you select three books from a shelf that contains five books? The answer to this question involves combinations. Keep in mind that if you select three books labeled A, B, and C, then any permutation of these three books is considered the same (the set A, B, and C and the set B, A, and C are considered the same selection).

As a third example, how many 5-digit binary numbers contain exactly three 1 values? The answer to this question also involves calculating a combinatorial value. In case you're wondering, the answer is `C(5,3) = 5!/[3! * 2!] = 10`, provided that we allow for leading zeroes. In fact, this is also the answer to the preceding question about selecting different subsets of books.

You can generalize the previous question by asking how many 4-digit, 5-digit, and 6-digit numbers contain exactly three 1s? The answer is the sum of these values (provided that leading zeroes are permitted):

```
C(4,3) + C(5,3) + C(6,3) = 4 + 10 + 20 = 34
```

Working With Permutations

Consider the following task: given 6 books, how many ways can you display them side by side? The possibilities are listed here:

position #1: 6 choices

position #2: 5 choices

position #3: 4 choices

position #4: 3 choices

position #5: 2 choices

position #6: 1 choice

The answer is 6x5x4x3x2x1 = 6! = 720. In general, if you have n books, there are n! different ways that you can order them (i.e., display them side by side).

Working With Combinations

Now let's look at a slightly different question: how many ways can you select three books from those six books? Here's the first approximation:

position #1: 6 choices

position #2: 5 choices

position #3: 4 choices

Since the number of books in any position is independent of the other positions, the first answer might be 6x5x4 = 120. However, this answer is incorrect because it includes different orderings of three books, but the sequence of books (A,B,C) is the same as (B,A,C) and every other recording of the letters A, B, and C.

As a concrete example, suppose that the books are labeled book #1, book #2, ..., book #6, and suppose that you select book #1, book #2, and book #3. Here is a list of all the different orderings of those three books:

```
123
132
213
231
312
321
```

The number of different orderings of 3 books is 3x2x1 = 3! = 6. However, from the standpoint of purely selecting three books, we must treat all six orderings as the same. Hence the correct answer is 6x5x4/[3x2x1] = 120/3! = 120/6 = 20.

Now let's multiply the numerator and the denominator by 3x2x1, which gives us this number: 6x5x4x3x2x1/[3x2x1 * 3x2x1] = 6!/[3! * 3!]

If we perform the preceding task of selecting three books from eight books instead of six books, we get this result:

8x7x6/[3x2x1] = 8x7x6x5x4x3x2x1/[3x2x1 * 5x4x3x2x1] = 8!/[3! * 5!]

Now suppose you select twelve books from a set of thirty books. The number of ways that this can be done is shown here:

```
30x29x28x...x19/[12x11x...x2x1]
= 30x29x28x...x19x18x17x16x...x2x1/[12x11x...x2x1 *
  18x17x16x...x2x1]
= 30!/[12! * 18!]
```

The general formula for calculating the number of ways to select `k` books from `n` books is `n!/[k! * (n-k)!]`, which is denoted by the term `C(n,k)`. Incidentally, if we replace k by n-k in the preceding formula we get this result:

```
n!/[(n-k)! * (n-(n-k))!] = n!/[(n-k)! * k)!] = C(n,k)
```

Notice that the left side of the preceding snippet equals `C(n,n-k)`, and therefore we have shown that `C(n,n-k) = C(n,k)`

THE NUMBER OF SUBSETS OF A FINITE SET

In the preceding section, if we allow k to vary from 0 to n inclusive, then we are effectively looking at all possible subsets of a set of n elements, and the number of such sets equals 2^n. We can derive the preceding result in two ways.

Solution #1.

The first way is the shortest explanation (and might seem like clever hand waving) and it involves visualizing a row of n books. In order to find every possible subset of those n books, we need only consider that there are two actions for the first position: either the book is selected or it is not selected.

Similarly, there are two actions for the second position: either the second book is selected or it is not selected. In fact, for every book in the set of n books there are the same two choices. Keeping in mind that the selection (or not) of a book in a given position is independent of the selection of the books in every other position, the number of possible choices equals:

```
2x2x...x2 (n times) = 2^n.
```

Solution #2.

Recall the following formulas from algebra:

```
(x+y)^2 = x^2 + 2*x*y + y^2
        = C(2,0)*x^2 + C(2,1)*x*y + C(2,2)*y^2
```

```
(x+y)^3 = x^3 + 4*x^2*y + 6*x^x*y^2 + 4*x*y^2 + y^3
        = C(3,0)*x^3 + C(3,0)*x^2*y + C(3,0)*x^x*y^2 +
          C(3,0)*x*y^2 + C(3,0)*y^3
```

In general, we have the following formula:

```
             n
(x+y)^n =   SUM C(n,k)*x^k*y^(n-k)
            k=0
```

Now set x=y=1 in the preceding formula and we get the following:

```
        n
2^n =  SUM C(n,k)
       k=0
```

The right-side of the preceding formula is the sum of the number of all possible subsets of a set of n elements, which the left side shows is equal to 2^n.

TASK: SUBSETS CONTAINING A VALUE LARGER THAN K

The more complete description of the task for this section is as follows: given a set N of numbers and a number k, find the number of subsets of N that contain at least one number that is larger than k. This *counting* task is an example of a coding task that can easily be solved as a combinatorial problem: you might be surprised to discover that the solution involves a single (and simple) line of code. Let's define the following set of variables:

```
N     = a set of numbers
|N|   = # of elements in N (= n)
NS    = the non-empty subsets of N
P(NS) = the number of non-empty subsets of N ( = |NS|)
M     = the numbers {n| n < k} where n is an element of N
|M|   = # of elements in M (= m)
MS    = the non-empty subsets of M
P(MS) = the number of non-empty subsets of M (= |MS|)
Q     = subsets of N that contain at least one number
        larger than k
```

Note that the set NS is partitioned into the sets Q and M, and that the union of Q and M is NS. In other words, a non-empty subset of N is either in Q or in M, but not in both. Therefore, the solution to the task can be expressed as: `|Q| = P(NS) - P(MS)`.

Moreover, the sets in M do not contain any number that is larger than k, which means that no element (i.e., subset) in M is an element of Q, and conversely, no element of Q is an element of M.

Recall from a previous result in this chapter that if a set contains m elements, then the number of subsets is `2**m`, and the number of *non-empty* subsets is `2**m - 1`. Hence, the answer for this task is `(2**n - 1) - (2**m - 1)`.

Listing 2.19 displays the contents of `subarrays_max_k.py` that calculates the sum of a set of binomial coefficients.

LISTING 2.19: SubarraysMaxK.java

```
// Time Complexity: O(1)
Public class SubarraysMaxK
{
   //##############################################
   // N = a set with n elements
   // M = a set with m elements
   //
   // N has 2^n - 1 non-empty subsets
   // M has 2^m - 1 non-empty subsets
   //
   // O = subsets of N with at least one element > k
   // P = subsets of N with all numbers <= k
   //
   // |P| = 2**m-1
   // and |O| = |N| - |P| = (2**n-1) - (2**m-1)
   //##############################################

   // number of subarrays whose maximum element > k
   public static int count_subsets(int n, int m) {
     //System.out.println("n = "+n+" m = "+m);
     int count = (int)(Math.pow(2,n) - 1) - (int)(Math.
pow(2,m) - 1);
     return count;
   }

   public static void main(String[] args)
   {
     //int[] arr = new int[]{ 1, 2, 3, 4, 5, 6, 7, 8 };
     int[] arr = new int[]{ 1, 2, 3, 4 };

     System.out.println("Array elements:");
     for (int num : arr)
        System.out.print(num+" ");
     System.out.println();

     int arr_len = arr.length;

     int[] arrk = new int[]{1, 2, 3, 4};
     for (int overk : arrk) {
       int count = count_subsets(arr_len, overk);

       System.out.println("overk:   "+overk);
       System.out.println("count:   "+count);
       System.out.println("---------------");
     }
   }
}
```

Listing 2.19 contains the Java code that implements the details that are described at the beginning of this section. Launch the code and you will see the following output:

```
Array elements:
1 2 3 4
overk:   1
count:   14
---------------
overk:   2
count:   12
---------------
overk:   3
count:   8
---------------
overk:   4
count:   0
--------------
```

Although the set N in Listing 2.19 contains a set of consecutive integers from 1 to n, the code works correctly for unsorted arrays or arrays that do not contain consecutive integers. In the latter case, you would need a code block to count the number of elements that are less than a given value of k.

SUMMARY

This chapter started with an introduction to recursion, along with various code samples that involve recursion, such as calculating factorial values, Fibonacci numbers, the sum of an arithmetic series, the sum of a geometric series, the GCD of a pair of positive integers, and the LCM of a pair of positive integers.

Moreover, you learned about concepts in combinatorics, and how to derive the formula for permutations and combinations of sets of objects. Finally, you saw two ways to show that the number of subsets of a set of n elements equals 2^n.

CHAPTER 3

Strings and Arrays

This chapter contains Java-based code samples that solve various tasks involving strings and arrays. The code samples in this chapter consist of the following sequence: examples that involve scalars and strings, followed by examples involving vectors (explained further at the end of this introduction), and then some examples involving 2D matrices. Note that the first half of Chapter 2 is relevant for the code samples in this chapter that involve recursion.

The first part of this chapter starts with a quick overview of the time complexity of algorithms, followed by various `Java` code samples, such as finding palindromes, reversing strings, and determining if the characters in a string unique.

The second part of this chapter discusses 2D arrays, along with `NumPy`-based code samples that illustrate various operations that can be performed on 2D matrices. This section also discusses 2D matrices, which are 2D arrays, along with some tasks that you can perform on them. This section also discusses multidimensional arrays, which have properties that are analogous to lower-dimensional arrays.

One other detail to keep in mind pertains to the terms vectors and arrays. In mathematics, a vector is a one-dimensional construct, whereas an array has at least two dimensions. In software development, an array can refer to a one-dimensional array or a higher-dimensional array (depending on the speaker). In this book a vector is always a one-dimensional construct. However, the term *array* always refers to a one-dimensional array; higher dimensional arrays will be referenced as "2D array," "3D array," and so forth. Therefore, the tasks involving 2D arrays start from the section titled "Working With 2D Arrays."

TIME AND SPACE COMPLEXITY

Algorithms are assessed in terms of the amount of space (based on input size) and the amount of time required for the algorithms to complete their execution, which is represented by "big O" notation. There are three types of time complexity: best case, average case, and worst case. Keep in mind that an algorithm with a very good "best case" performance can have a relatively poor worse case performance.

Recall that `O(n)` means that an algorithm executes in linear time because its complexity is bounded above and below by a linear function. For example, if three algorithms require `2*n`, `5*n`, or `n/2` operations, respectively, then all of them have O(n) complexity.

Moreover, if the best, average, and worst time performance for a linear search is 1, n/2, and n operations, respectively, then those operations have O(1), O(n), and O(n), respectively. In general, if there are two solutions T1 and T2 for a given task such T2 is more efficient than T1, then T2 requires either less time or less memory. For example, if T1 is an iterative solution for calculating Fibonacci numbers and T2 involves a recursive solution, then T1 is more efficient than T2 in terms of time, but T1 also requires an extra array to store intermediate values.

The *time-space trade-off* refers to reducing either the amount of time or the amount of memory that is required for executing an algorithm, which involves choosing one of the following:

- execute in less time and more memory
- execute in more time and less memory

Although reducing both time and memory is desirable, it's also a more challenging task. Another point to keep in mind is the following inequalities (logarithms can be in any base that is greater than or equal to 2) for any positive integer `n` > 1:

```
O(log n) < O(n) < O(n*log n) < O(n^2)
```

In addition, the following inequalities with powers of `n`, powers of 2, and factorial values are also true:

```
O(n**2) < O(n**3) < O(2**n) < O(n!)
```

If you are unsure about any of the preceding inequalities, perform an online search for tutorials that provide the necessary details.

TASK: MAXIMUM AND MINIMUM POWERS OF AN INTEGER

The code sample in this section shows you how to calculate the largest (smallest) power of a number `num` whose base is `k` that is less than (greater than) `num`, where `num` and `k` are both positive integers.

For example, 16 is the largest power of 2 that is *less* than 24 and 32 is the smallest power of 2 that is *greater* than 24. As another example, 625 is the largest power of 5 that is *less* than 1000 and 3125 is the smallest power of 5 that is *greater* than 1000.

Listing 3.1 displays the contents of `MaxMinPowerk2.java` that illustrates how to calculate the largest (smallest) power of a number whose base is `k` that is less than (greater than) a given number. Just to be sure that the task is clear: `num` and `k` are positive integers, and the purpose of this task is two-fold:

- find the *largest* number `powk` such that `k**powk <= num`
- find the *smallest* number `powk` such that `k**powk >= num`

LISTING 3.1: MaxMinPowerk2.java

```
public class MaxMinPowerk2
{
   public static int[] MaxMinPowerk(int num, int k)
   {
      int powk = 1;
      while(powk <= num)
      {
        powk *= k;
      }

      if(powk > num) powk /= k;

      return new int[]{(int)powk, (int)powk*k};
   }

   public static void main(String[] args)
   {
      int lowerk, upperk;
      int[] nums = new int[]{24,17,1000};
      int[] powers = new int[]{2,3,4,5};

      for(int num : nums)
      {
        for(int k : powers)
        {
          int[] results = max_min_powerk(num, k);
          lowerk = results[0];
          upperk = results[1];
          System.out.println("num: "+num+" lower "+lowerk+"
          upper: "+upperk);
        }

        System.out.println();
      }
   }
}
```

Listing 3.1 starts with the function `MaxMinPowerk()` that contains a loop that repeatedly multiplies the local variable `powk` (initialized with the value 1) by `k`. When `powk` exceeds the parameter `num`, then `powk` is divided by `k` so that we have the lower bound solution.

Note that this function returns `powk` *and* `powk*k` because this pair of numbers is the lower bound and higher bound solutions for this task. Launch the code in Listing 3.1 and you will see the following output:

```
num: 24 lower 16 upper: 32
num: 24 lower 9 upper: 27
num: 24 lower 16 upper: 64
num: 24 lower 5 upper: 25

num: 17 lower 16 upper: 32
num: 17 lower 9 upper: 27
num: 17 lower 16 upper: 64
num: 17 lower 5 upper: 25

num: 1000 lower 512 upper: 1024
num: 1000 lower 729 upper: 2187
num: 1000 lower 256 upper: 1024
num: 1000 lower 625 upper: 3125
```

TASK: BINARY SUBSTRINGS OF A NUMBER

Listing 3.2 displays the contents of the `BinaryNumbers.java` that illustrates how to display all binary substrings whose length is less than or equal to a given number.

LISTING 3.2: BinaryNumbers.java

```
public class BinaryNumbers
{
   public static void binaryValues(int width)
   {
      System.out.println("Binary values of width "+width+":");

      String bin_value = "";
      int power = (int)java.lang.Math.pow(2,width);

      for(int i=0; i<power; i++)
      {
         bin_value = Integer.toBinaryString(i);
         System.out.println(bin_value);
      }
      System.out.println("");
   }

   public static void main(String[] args)
   {
      int max_width = 4;
      for (int i=1; i<max_width; i++) {
```

```
            //System.out.print(el + " ");
            binaryValues(i);
        }
    }
}
```

Listing 3.2 starts with the method `binaryValues()` whose loop iterates from 0 to `2**width`, where `width` is the parameter for this function. The loop variable is `i` and during each iteration, `bin_value` is initialized with the binary value of `i`.

Next, the variable `str_value` is the string-based value of `bin_value`, which is stripped of the two leading characters `0b`. Launch the code in Listing 3.2 and you will see the following output:

```
Binary values of width 1:
0
1

Binary values of width 2:
0
1
10
11

Binary values of width 3:
0
1
10
11
100
101
110
111
```

TASK: COMMON SUBSTRING OF TWO BINARY NUMBERS

Listing 3.3 displays the contents of `CommonBits.java` that illustrates how to find the longest common substring of two binary strings.

LISTING 3.3: CommonBits.java

```
public class CommonBits
{
   public static int commonBits(int num1, int num2)
   {
      String bin_str1 = Integer.toBinaryString(num1);
      String bin_str2 = Integer.toBinaryString(num2);

      if(bin_str2.length() < bin_str1.length())
```

```
      {
        while(bin_str2.length() < bin_str1.length())
          bin_str2 = "0" + bin_str2;
      }

      if(bin_str1.length() < bin_str2.length())
      {
        while(bin_str1.length() < bin_str2.length())
          bin_str1 = "0" + bin_str1;
      }

      System.out.println(num1+"="+bin_str1);
      System.out.println(num2+"="+bin_str2);

      int common = 0;
      String common_bits2 = "";
      for(int i=0; i<bin_str1.length(); i++)
      {
        if((bin_str1.charAt(i) == bin_str2.charAt(i)) &&
           (bin_str1.charAt(i) =='1'))
        {
           common_bits2 = common_bits2 + "1";
           common++;
        }
      }
      return common;
   }

   public static void main(String[] args)
   {
      int[] nums1 = new int[]{61,28, 7,100,189};
      int[] nums2 = new int[]{51,14,28,110, 14};

      for(int idx=0; idx<nums1.length; idx++)
      {
        int num1 = nums1[idx];
        int num2 = nums2[idx];
        int common = commonBits(num1, num2);

        System.out.println(
           num1+" and "+num2+" have "+common+" bits in
           common\n");
      }
   }
}
```

Listing 3.3 starts with the function `commonBits()` that initializes the binary strings `bin_str1` and `bin_str2` with the binary values of the two input parameters. The next two loops left-pad the strings `bin_str1` and `bin_str2` to ensure that they have the same number of digits.

The next portion of Listing 3.3 is a loop iterates from 0 to the length of the string `bin_str1` in order to compare each digit to the corresponding digit in bit_str2 to see if they are both equal and also equal to 1. Each time that the digit 1 is found, the value of `common` (initialized with the value 0) is incremented. When the loop terminates, the variable `common` equals the number of common bits in `bin_num1` and `bin_num2`.

The final portion of Listing 3.3 is the main() method that initializes two arrays with integer values, followed by a loop that iterates through these two arrays in order to count the number of common bits in each pair of numbers. Launch the code in Listing 3.3 and you will see the following output:

```
61=111101
51=110011
61 and 51 have 3 bits in common

28=11100
14=01110
28 and 14 have 2 bits in common

7=00111
28=11100
7 and 28 have 1 bits in common

100=1100100
110=1101110
100 and 110 have 3 bits in common

189=10111101
14=00001110
189 and 14 have 2 bits in common
```

TASK: MULTIPLY AND DIVIDE VIA RECURSION

Listing 3.4 displays the contents of `RecursiveMultiply.java` that illustrates how to compute the product of two positive integers via recursion.

LISTING 3.4: RecursiveMultiply.java

```
class RecursiveMultiply
{
   public static int addRepeat(int num, int times, int sum)
   {
      if(times == 0)
        return sum;

      return addRepeat(num, times-1, num+sum);
   }
```

```
    public static void main(String args[])
    {
       int[] arr1 = {9,13,25,17,100};
       int[] arr2 = {5,10,25,10,100};

       for(int i=0; i<arr1.length; i++) {
          int num1 = arr1[i];
          int num2 = arr2[i];
          int prod = addRepeat(num1, num2, 0);
          System.out.println("remainder of "+num1+"/"+num2+"
          = "+prod);
       }
    }
}
```

Listing 3.4 starts with the method `addRepeat(num,times,sum)` that performs repeated addition by recursively invokes itself. Note that this method uses tail recursion: each invocation of the method replaces times with `times-1` and also replaces `sum` with `num+sum` (the latter is the tail recursion). The terminating condition is when `times` equals 0, at which point the method returns the value of `sum`. Launch the code in Listing 3.4 and you will see the following output:

```
product of 5 and 9 = 45
product of 13 and 10 = 130
product of 25 and 25 = 625
product of 17 and 10 = 170
product of 100 and 100 = 10000
```

Listing 3.5 displays the contents of the `RecursiveDivide.java` that illustrates how to compute the quotient of two positive integers via recursion.

LISTING 3.5: RecursiveDivide.java

```
class RecursiveDivide
{
   public static int subRepeat(int num1, int num2, int remainder)
   {
      if(num1 < num2)
        return num1;
      else
        //print("num1-num2:",num1-num2,"num2:",num2)
        return sub_repeat(num1-num2, num2, remainder);
   }

   public static void main(String args[])
   {
      int[] arr1 = {9,13,25,17,100};
      int[] arr2 = {5,10,25,10,100};

      for(int i=0; i<arr1.length; i++) {
```

```
            int num1 = arr1[i];
            int num2 = arr2[i];
            int prod = subRepeat(num1, num2, 0);
            System.out.println("remainder of "+num1+"/"+num2+"
            = "+prod);
         }
      }
   }
```

Listing 3.5 contains code that is very similar to Listing 3.4: the difference involves replacing addition with subtraction. Launch the code in Listing 3.5 and you will see the following output:

```
remainder of 9 / 5 = 4
remainder of 13 / 10 = 3
remainder of 25 / 25 = 0
remainder of 17 / 10 = 7
remainder of 100 / 100 = 0
```

TASK: SUM OF PRIME AND COMPOSITE NUMBERS

Listing 3.6 displays the contents of the `PairSumSorted.java` that illustrates how to determine whether or not a sorted array contains the sum of two specified numbers.

LISTING 3.6: PairSumSorted.java

```
//given two numbers num1 and num2 in a sorted array,
//determine whether or not num1+num2 is in the array

public class PairSumSorted
{
   public static void checkSum(int[] arr1,int num1,int num2)
   {
      int ndx1 = 0;
      int ndx2 = arr1.length-1;
      System.out.println("Anum1: "+num1+" num2: "+num2);

      while(arr1[ndx1] < num1)
        ndx1 += 1;
      System.out.println("Bndx1: "+ndx1+" ndx2: "+ndx2);

      int sum = num1+num2;
      while(arr1[ndx2] > num2)
        ndx2 -= 1;

      System.out.print("Array: ");
      for(int i=0; i< arr1.length; i++) {
        System.out.print(arr1[i]+" ");
      }
      System.out.println("");
```

```
        System.out.println("Cndx1: "+ndx1+" ndx2: "+ndx2);
        System.out.println("num1:  "+num1+" num2: "+num2);
        System.out.println("ndx1:  "+ndx1+" ndx2: "+ndx2);
        /* NOTE: arr1[ndx1] >= num1 AND arr1[ndx2] >= sum */

        if(arr1[ndx1]+arr1[ndx2] == sum)
          System.out.println(
            "=> FOUND sum "+sum+" for ndx1 = "+ndx1+" ndx2 =
            "+ndx2);
        else
          System.out.println("=> NOT FOUND "+sum);
        System.out.println();
    }

    public static void main(String args[])
    {
        int[] arr1 = { 20,50,100,120,150,200,250,300 };
        int num1 = 60;
        int num2 = 90;
        checkSum(arr1,num1,num2);

        num1 = 60;
        num2 = 100;
        checkSum(arr1,num1,num2);

        int[] arr2 = { 3,3 };
        num1 = 3;
        num2 = 3;
        checkSum(arr2,num1,num2);
    }
}
```

Listing 3.6 starts with the function `is_prime()` that determines whether or not its input parameter is a prime number. The next portion of code in Listing 3.6 is a loop that ranges from 0 to the number of elements. During each iteration, the current number is added to the variable `prime_sum` if that number is a prime; otherwise, it is added to the variable `comp_sum`.

The final portion of Listing 3.6 displays the sum of the even numbers and the sum of the odd numbers in the input array `arr1`. Launch the code in Listing 3.6 and you will see the following output:

```
prime list: [ 5. 17. 23. 47.]
comp  list: [10. 30. 50.]
prime sum:  92
comp sum:   90
```

The next portion of this chapter contains various examples of string-related tasks. If need be, you can review the relevant portion of Chapter 1 regarding some of the `Java` built-in string functions, such as `int()` and `len()`.

TASK: COUNT WORD FREQUENCIES

Listing 3.7 displays the contents of the `WordFrequency.java` that illustrates how to determine the frequency of each word in an array of sentences.

LISTING 3.7: WordFrequency.java

```
import java.util.ArrayList;
import java.util.Arrays;
import java.util.Hashtable;
import java.util.Enumeration;

public class WordFrequency
{
   public static void main(String[] args)
   {
      String[] sent1 =
        new String[]{"I", "love", "thick", "pizza"};
      String[] sent2 =
        new String[]{"I", "love", "deep", "dish","pizza"};
      String[] sent3 =
        new String[]{"Pepperoni","and","sausage","pizza"};
      String[] sent4 =
        new String[]{"Pizza", "with", "mozzarrella"};

      ArrayList<String> sents =
        new ArrayList<String>(Arrays.asList(sent1));
      sents.addAll(Arrays.asList(sent2));
      sents.addAll(Arrays.asList(sent3));
      sents.addAll(Arrays.asList(sent4));

      ArrayList<String> words = new ArrayList<String>();
      ArrayList<String> tokens = new ArrayList<String>();

      Hashtable<String,Integer> ht = new
      Hashtable<String,Integer>();

      for(int idx=0; idx<sents.size(); idx++)
      {
         String word = sents.get(idx);
         tokens.add(word);

         // remove this snippet if you want case sensitivity:
         word = word.toLowerCase();

         if (!ht.contains(word)) {
           ht.put(word,0);
         }
         int count = ht.get(word);
         ht.put(word,count+1);
      }
```

```
      //System.out.println("tokens: "+tokens);

      Enumeration enumer = ht.keys();

      System.out.println("Words and their frequency:");
      while (enumer.hasMoreElements()) {
         String key = (String)enumer.nextElement();
         System.out.println("key: "+key+" occurrences:
         "+ht.get(key));
      }
   }
}
```

Listing 3.7 starts with the `main()` method that that initializes four variables with a list of strings, followed by a code block that initializes the variable sents as an ArrayList that contains sent1, sent2, sent3, and sent4.

The next portion of Listing 3.7 initializes words and tokens as instances of ArrayList, followed by the initialization of the variable ht as an instance of the Hashtable class.

Next, a loop iterates through each "sentence" in sents, and then appends each "word" (i.e., string) to tokens, followed by converting "word" to lower-case, whose value in ht is updated appropriately.

The final portion of Listing 3.7 iterates through the keys of ht in order to print the number of occurrences of each key in ht, which equals the frequency of each word in the initial set of sentences. Launch the code in Listing 3.7 and you will see the following output:

```
Words and their frequency:
key: dish occurrences: 1
key: pepperoni occurrences: 1
key: i occurrences: 1
key: and occurrences: 1
key: thick occurrences: 1
key: love occurrences: 1
key: sausage occurrences: 1
key: pizza occurrences: 1
key: deep occurrences: 1
key: with occurrences: 1
key: mozzarrella occurrences: 1
```

TASK: CHECK IF A STRING CONTAINS UNIQUE CHARACTERS

The solution involves keeping track of the number of occurrences of each `ASCII` character in a string, and returning `False` if that number is greater than 1 for any character (otherwise return `True`). Hence, one constraint for this solution is that it's restricted to Indo-European languages that do not have accent marks.

Listing 3.8 displays the contents of `UniqueChars.java` that illustrates how to determine whether or not a string contains unique letters.

LISTING 3.8: UniqueChars.java

```
//128 characters for ASCII and 256 characters for extended
ASCII
public class UniqueChars
{
   public static Boolean uniqueChars(String str)
   {
      boolean result = false;
      if (str.length() > 128)
        return result;

      str = str.toLowerCase();

      int[] char1_set = new int[128];

      for(int idx=0; idx<str.length(); idx++)
      {
        char char1 = str.charAt(idx);
        int val = 'z' - char1; // # error
        //System.out.println("val: "+val);

        if (char1_set[val] == 1)
          return result;
        else
          char1_set[val] = 1;
      }

      return !result;
   }

   public static void main(String[] args)
   {
      String[] arr1 =
       new String[]{"a string", "second string", "hello world"};

      for(String str : arr1)
      {
         System.out.println("string: "+str);
         boolean result = uniqueChars(str);
         System.out.println("unique: "+result);
         System.out.println();
      }
   }
}
```

Listing 3.8 starts with the function `unique_chars()` that converts its parameter `str` to lower case letters and then initializes the 1x128 integer array `char_set` whose values are all 0. The next portion of this function iterates through the characters of the string `str` and initializes the integer variable `val` with the offset position of each character from the character `z`.

If this position in `char_set` equals 1, then a duplicate character has been found; otherwise, this position is initialized with the value 1. Note that the value `False` is returned if the string `str` contains duplicate letters, whereas the value `True` is returned if the string str contains unique characters. Launch the code in Listing 3.8 and you will see the following output:

```
string: a string
unique: True

string: second string
unique: False

string: hello world
unique: False
```

TASK: INSERT CHARACTERS IN A STRING

Listing 3.9 displays the contents of `InsertChars.java` that illustrates how to insert each character of one string in every position of another string.

LISTING 3.9: InsertChars.java

```
import java.io.*;
import java.util.*;

public class InsertChars
{
   public static String insertChar(String str1, char chr)
   {
     String inserted="", left="", right="", result = str1;

     result = chr + str1;
     for(int i=0;i<str1.length(); i++)
     {
        left  = str1.substring(0,i+1);
        right = str1.substring(i+1);
        inserted = left + chr + right;
     }

     result = result + " " + inserted;
     return result;
   }

   public static void main(String[] args)
   {
      String str1 = "abc";
      String str2 = "def";
      System.out.println("str1: "+str1);
      System.out.println("str2: "+str2);
```

```
        String newStr="", insertions = "";
        for(int i=0; i<str2.length(); i++)
        {
           newStr = insertChar(str1, str2.charAt(i));
           insertions = insertions+ " " + newStr;
        }

        System.out.println("result:"+insertions);
    }
}
```

Listing 3.9 starts with the method `insertChar()` that has a string `str1` and a character `chr` as input parameters. The next portion of code is a loop whose loop variable is `i`, which is used to split the string `str1` into two strings: the left substring from positions 0 to `i`, and the right substring from position `i+1`. A new string with three components is constructed: the left string, the character `chr`, and the right string.

The next portion of Listing 3.9 contains a loop that iterates through each character of `str2`; during each iteration, the code invokes `insert_char()` with string `str1` and the current character. The number of new strings generated by this code equals the following product: `(len(str1)+1)*len(str2)`.

Launch the code in Listing 3.9 and you will see the following output:

```
str1: abc
str2: def
result:  dabc adbc abdc abcd eabc aebc abec abce fabc afbc
abfc abcf
```

TASK: STRING PERMUTATIONS

There are several ways to determine whether or not two strings are permutations of each other. One way involves sorting the strings alphabetically: if the resulting strings are equal, then they are permutations of each other.

A second technique is to determine whether or not they have the same number of occurrences for each character. A third way is to add the numeric counterpart of each letter in the string: if the numbers are equal and the strings have the same length, then they are permutations of each other.

Listing 3.10 displays the contents of `StringPermute.java` that illustrates how to determine whether or not two strings are permutations of each other.

LISTING 3.10: StringPermute.java

```
import java.util.Arrays;

public class StringPermute
{
   public static void permute(String str1,String str2)
```

```
      {
         char temp1[] = str1.toCharArray();
         char temp2[] = str2.toCharArray();
         Arrays.sort(temp1);
         Arrays.sort(temp2);

         String str1d = new String(temp1);
         String str2d = new String(temp1);
         Boolean permute = (str1d == str2d);

         System.out.println("string1: "+str1);
         System.out.println("string2: "+str2);
         System.out.println("permuted:"+permute);
         System.out.println();
      }

      public static void main(String[] args)
      {
         String[] strings1 = new String[]{"abcdef", "abcdef"};
         String[] strings2 = new String[]{"efabcf", "defabc"};

         for(int idx=0; idx<strings1.length; idx++) {
            String str1 = strings1[idx];
            String str2 = strings2[idx];
            permute(str1,str2);
         }
      }
   }
```

Listing 3.10 starts with the function `permute()` that takes the two strings `str1` and `str2` as parameters. Next, the strings `str1d` and `str2d` are initialized with the result of sorting the characters in the strings `str1` and `str2`, respectively. At this point, we can determine whether or not `str1` and `str2` are permutations of each other by determining whether or not the two strings `str1d` and `str2d` are equal. Launch the code in Listing 3.10 and you will see the following output:

```
string1:  abcdef
string2:  efabcf
permuted: False

string1:  abcdef
string2:  defabc
permuted: True
```

TASK: CHECK FOR PALINDROMES

One way to determine whether or not a string is a palindrome is to compare the string with the reverse of the string: if the two strings are equal, then the string is a palindrome. Moreover, there are two ways to reverse a string: one way involves the `Java reverse()` function, and another way is to process the characters in the given string in a right-to-left fashion, and to append each character to a new string.

Another technique involves iterating through the characters in a left-to-right fashion and comparing each character with its corresponding character that is based on iterating through the string in a right-to-left fashion.

Listing 3.11 displays the contents of the `Palindrome1.java` that illustrates how to determine whether or not a string or a positive integer is a palindrome.

LISTING 3.11: Palindrome1.java

```
public class Palindrome1
{
   public static Boolean palindrome1(String str) {
     int full_len = str.length();
     int half_len = str.length()/2;

     for(int i=0; i<half_len; i++) {
       char lchar = str.charAt(i);
       char rchar = str.charAt(full_len-1-i);
       if(lchar != rchar)
         return false;
     }
     return true;
   }

   public static void main(String[] args)
   {
      String[] arr1 = new String[]{"rotor", "tomato",
      "radar","maam"};
      String[] arr2 = new String[]{"123", "12321", "555"};

      // CHECK FOR STRING PALINDROMES:
      for(String str : arr1)
      {
        Boolean result = palindrome1(str);
        System.out.println("String: "+str+ " palindrome:
        "+result);
      }

      // CHECK FOR NUMERIC PALINDROMES:
      for(String num : arr2)
      {
        System.out.print("Number: "+num);
        String str1 = num;
        String str2 = "";
        for(int i=0; i<str1.length(); i++) {
          str2 = str2 + str1.charAt(i);
        }
```

```
        Boolean result = palindrome1(str2);
        System.out.println(" palindrome:  "+result);
      }
   }
}
```

Listing 3.11 starts with the function `palindrome1()` with parameter `str` that is a string. This function contains a loop that starts by comparing the left-most character with the right-most character of the string `str`. The next iteration of the loop advances to the second position of the left side of the string and compares that character with the character whose position is second from the right end of the string. This step-by-step comparison continues until the middle of the string is reached. During each iteration of the loop, the value `False` is returned if the pair of characters is different. If all pairs of characters are equal, then the string must be a palindrome, in which case the value `True` is returned.

The next portion of Listing 3.11 contains an array `arr1` of strings and an array `arr2` of positive integers. Next, another loop iterates through the elements of `arr1` and invokes the `palindrome1` function to determine whether or not the current element of `arr1` is a palindrome. Similarly, a second loop iterates through the elements of `arr2` and invokes the `palindrome1` function to determine whether or not the current element of `arr2` is a palindrome. Launch the code in Listing 3.11 and you will see the following output:

```
String: rotor palindrome: true
String: tomato palindrome: false
String: radar palindrome: true
String: maam palindrome: true
Number: 123 palindrome:  false
Number: 12321 palindrome:  true
Number: 555 palindrome:  true
```

TASK: CHECK FOR THE LONGEST PALINDROME

This section extends the code in the previous section by examining substrings of a given string. Listing 3.12 displays the contents of `LongestPalindrome.java` that illustrates how to determine the longest palindrome in a given string. Note that a single character is always a palindrome, which means that every string has a substring that is a palindrome (in fact, any single character in any string is a palindrome).

LISTING 3.12: LongestPalindrome.java

```
public class LongestPalindrome1
{
   public static String checkString(String str)
   {
     int result = 0; // 0 = palindrome 1 = not palindrome
```

```
    int str_len  = str.length();
    int str_len2 = str.length()/2;

    for(int i=0; i<str_len2; i++) {
      if(str.charAt(i) != str.charAt(str_len-i-1)) {
        result = 1;
        break;
      }
    }

    if(result == 0)
      return str;
    else
      return "";
  }

  public static Boolean palindrome1(String str)
  {
    int full_len = str.length();
    int half_len = str.length()/2;

    for(int i=0; i<half_len; i++) {
      char lchar = str.charAt(i);
      char rchar = str.charAt(full_len-1-i);
      if(lchar != rchar)
        return false;
    }
    return true;
  }

  public static void main(String[] args)
  {
     String[] my_strings = new String[]
         {"abc","abb","abccba","azaaza","abcdefgabccbax"};

     String max_pal_str = "";
     int max_pal_len = 0;
     String sub_str = "";

     for(String my_str : my_strings) {
       max_pal_str = "";
       max_pal_len = 0;
       int my_str_len = my_str.length();

       for(int i=0; i<my_str.length()-1; i++) {
          for(int j=1; j<my_str.length()-i+1; j++) {
             sub_str = my_str.substring(i,i+j);
             String a_str = checkString(sub_str);

             if(a_str != "") {
                if(max_pal_len < a_str.length()) {
                   max_pal_len = a_str.length();
```

```
                        max_pal_str = a_str;
                    }
                }
            }
        }
        System.out.println("string: "+my_str);
        System.out.println("maxpal: "+max_pal_str);
        System.out.println();
    }
  }
}
```

Listing 3.12 contains logic that is very similar to Listing 3.11. However, the main difference is that there is a loop that checks if *substrings* of a given string are palindromes. The code also keeps track of the longest palindrome and then prints its value and its length when the loop finishes execution.

Note that it's possible for a string to contain multiple palindromes of maximal length: the code in Listing 3.12 finds only the first such palindrome. However, it might be a good exercise to modify the code in Listing 3.12 to find all palindromes of maximal length. Launch the code in Listing 3.12 and you will see the following output:

```
string: abc
maxpal: a

string: abb
maxpal: bb

string: abccba
maxpal: abccba

string: azaaza
maxpal: azaaza

string: abcdefgabccbax
maxpal: abccba
```

WORKING WITH SEQUENCES OF STRINGS

This section contains `Java` code samples that search strings to determine the following:

- the maximum length of a sequence of consecutive 1s in a string
- a given sequence of characters in a string
- the maximum length of a sequence of unique characters

After you complete this section, you can explore variations of these tasks that you can solve using the code samples in this section.

The Maximum Length of a Repeated Character in a String

Listing 3.13 displays the contents of `MaxCharSequence.java` that illustrates how to find the maximal length of a repeated character in a string.

LISTING 3.13: MaxCharSequence.java

```
public class MaxCharSequence
{
   public static void maxSeq(String my_str,char ch)
   {
      int max = 0, left = 0, right = 0, counter = 0;
      for(int i=0; i<my_str.length(); i++) {
        char curr_ch = my_str.charAt(i);

        if(curr_ch == ch) {
          counter += 1;
          right = i;
          if(max < counter)
            max = counter;
            //System.out.println("new max:",max)
        } else {
          counter = 0;
          left = i;
          right = i;
        }
      }

      System.out.println("my_str: "+my_str);
      System.out.println("max sequence of "+ch+": "+max);
      System.out.println();
   }

   public static void main(String[] args)
   {
      String[] str_list  = new String[]
                  {"abcdef","aaaxyz","abcdeeefghij"};

      String[] char_list = new String[]{"a","a","e"};

      for(int idx=0; idx<str_list.length; idx++) {
         String my_str = str_list[idx];
         String str    = char_list[idx];
         char ch       = char_list[idx].charAt(0);
         maxSeq(my_str,ch);
      }
   }
}
```

Listing 3.13 starts with the method `maxSeq()` whose parameters are a string `my_str` and a character `char`. This function contains a loop that iterates through each character of my_str and performs a comparison with char. As long as the characters equal char, the value of the variables right and counter are incremented: right represents the right-most index and counter contains the length of the substring containing the same character.

However, if a character in my_str differs from char, then counter is reset to 0, and left is reset to the value of right, and the comparison process begins anew. When the loop has completed execution, the variable counter equals the length of the longest substring consisting of equal characters.

The next portion of Listing 3.13 initializes the array str_list that contains a list of strings and the array char_list with a list of characters. The final loop iterates through the elements of str_list and invokes the function max_seq() with the current string and the corresponding character in the array char_list. Now launch the code in Listing 3.13 and you will see the following output:

```
my_str: abcdef
max sequence of a: 1

my_str: aaaxyz
max sequence of a: 3

my_str: abcdeeefghij
max sequence of e: 3
```

Find a Given Sequence of Characters in a String

Listing 3.14 displays the contents of MaxSubstrSequence.java that illustrates how to find the right-most substring that matches a given string.

LISTING 3.14: MaxSubstrSequence.java

```
public class MaxSubstrSequence
{
   public static void rightmostSubstr(String my_str,String
   substr)
   {
      int left = -1;
      int len_substr = substr.length();
      System.out.println("initial substr: "+substr);

      // check for substr from right to left:
      for(int i=my_str.length()-len_substr; i>=0; i--) {
        //String curr_str = my_str[i:i+len_substr];
        String curr_str = my_str.substring(i,i+len_substr);

        if(substr.equals(curr_str)) {
          left = i;
```

```
          break;
        }
      }

      if(left >= 0)
        System.out.println(
           substr+" is in index "+left+" of: "+my_str);
      else
        System.out.println(
           substr+" does not appear in "+my_str);
      System.out.println();
   }

   public static void main(String[] args)
   {
      String[] str_list    =
              new String[]{"abcdef","aaaxyz","abcdeeefghij"};

      String[] substr_list = new String[]{"bcd","aaa","cde"};

      for(int idx=0; idx<str_list.length; idx++) {
        String my_str = str_list[idx];
        String substr = substr_list[idx];
        rightmostSubstr(my_str,substr);
      }
   }
}
```

Listing 3.14 starts with the method `rightmostSubstr()` whose parameters are a string `my_str` and a substring `sub_str`. This function contains a loop that performs a right-most comparison of `my_str` and `sub_str`, and iteratively moves leftward one position until the loop reaches the first index position of the string `my_str`.

After the loop has completed its execution, the variable `left` contains the index position at which there is a match between `my_str` and `sub_str`, and its value will be nonnegative. If there is no matching substring, then the variable `left` will retain its initial value of -1. In either case, the appropriate message is printed. Launch the code in Listing 3.14 and you will see the following output:

```
initial substr: bcd
bcd is in index 1 of: abcdef

initial substr: aaa
aaa is in index 0 of: aaaxyz

initial substr: cde
cde is in index 2 of: abcdeeefghij
```

TASK: THE LONGEST SEQUENCES OF SUBSTRINGS

This section contains Java code samples that search strings to determine the following:

- the longest subsequence of unique characters in a given string
- the longest subsequence that is repeated in a given string

After you complete this section, you can explore variations of these tasks that you can solve using the code samples in this section.

The Longest Sequence of Unique Characters

Listing 3.15 displays the contents of LongestUnique.java that illustrates how to find the longest sequence of unique characters in a string.

LISTING 3.15: LongestUnique.java

```
import java.util.Hashtable;

public class LongestUnique
{
   public static void rightmostSubstr(String my_str)
   {
      int left = 0, right = 0;
      String sub_str = "", longest = "";
      Hashtable<String, Integer> ht = new
      Hashtable<String,Integer>();

      for(int pos=0; pos<my_str.length(); pos++) {
        String ch_str = "" + my_str.charAt(pos);

        Integer count = 0;
        if (ht.containsKey(ch_str))
          count = ht.get(ch_str);

        if (count > 0) {
          ht = new Hashtable<String,Integer>();
          left = pos+1;
          right = pos+1;
        } else {
          ht.put(ch_str, 1);

          String unique = my_str.substring(left,pos+1);
          //System.out.println("unique string: "+unique);

          if(longest.length() < unique.length()) {
            longest = unique;
            right = pos;
          } else {
```

```
            //System.out.println("new hashtable:");
            ht = new Hashtable<String,Integer>();
            left = pos+1;
            right = pos+1;
          }
        }
      }

      System.out.println("original string: "+my_str);
      System.out.println("longest unique:  "+longest);
   }

   public static void main(String[] args)
   {
      //String [] str_list = new String[]{"ACCC"};
      String [] str_list =
          new String[]{"abcdef","aaaxyz","abcdeeefghij"};

      for(int idx=0; idx<str_list.length; idx++) {
        String my_str = str_list[idx];
        rightmostSubstr(my_str);
      }
   }
}
```

Listing 3.15 starts with the method `rightmostSubstr()` whose parameter is a string `my_str`. This function contains a right-to-left loop and stores the character in the current index position in the dictionary `my_dict`. If the character has already been encountered, then it's a duplicate character, at which point the length of the current substring is compared with the length of the longest substring that has been found thus far, at which point the variable `longest` is updated with the new value. In addition, the left position `left` and the right position `right` are reset to `pos+1`, and the search for a unique substring begins anew.

After the loop has completed its execution, the value of the variable `longest` equals the length of the longest substring of unique characters.

The next portion of Listing 3.15 initializes the variable `str_list` as an array of strings, followed by a loop that iterates through the elements of `str_list`. The method `rightmostSubstr()` is invoked during each iteration in order to find the longest unique substring of the current string. Launch the code in Listing 3.15 and you will see the following output:

```
checking: abcdef
longest unique: abcdef

checking: aaaxyz
longest unique: axyz
```

```
checking: abcdeeefghij
longest unique: efghij
```

The Longest Repeated Substring

Listing 3.16 displays the contents of `MaxRepeatedSubstr.java` that illustrates how to find the longest substring that is repeated in a given string.

LISTING 3.16: MaxRepeatedSubstr.java

```
public class MaxRepeatedSubstr
{
   public static String[] checkString(String my_str, String
   sub_str, int pos)
   {
     String match = "", part_str = "";
     int left = 0, right = 0;

     int str_len = my_str.length();
     int sub_len = sub_str.length();
     //System.out.println("my_str: "+my_str+" sub_str: "+sub_str);

     for(int i=0; i<str_len-sub_len-pos; i++) {
       left  = pos+sub_len+i;
       right = left+sub_len;
       //System.out.println("left: "+left+" right: "+right);

       if(right > str_len) {
         //System.out.println("right too large: "+right);
         break;
       }

       part_str = my_str.substring(left,right);

       if(part_str.equals(sub_str)) {
         match = part_str;
         break;
       }
     }

     return new String[]{match,Integer.toString(left)};
   }

   public static void main(String[] args)
   {
      System.out.println("==> Repeating substrings have
      length >= 2");
      String [] my_strings =
       new String[]{"abc","abab","abccba","azaaza","abcdefg
       abccbaxyz"};
```

```
        for(String my_str : my_strings) {
          System.out.println("=> Checking current string:
          "+my_str);
          int half_len = (int)(my_str.length()/2);
          int max_len = 0;
          String max_str = "";

          for(int i=0; i<half_len+1; i++) {
            for(int j=2; j<half_len+1; j++) {
              String sub_str = my_str.substring(i,i+j);
              String[] result = checkString(my_str, sub_
              str,i);
              String a_str = result[0];
              int left     = Integer.parseInt(result[1]);

              if(a_str != "") {
               if(max_len < a_str.length()) {
                  max_len = a_str.length();
                  max_str = a_str;
                }
              }
            }
          }

          if(max_str != "")
            System.out.println("=> Maximum repeating
            substring: "+max_str+"\n");
          else
            System.out.println("No maximum repeating
            substring: "+my_str+"\n");
        }
    }
}
```

Listing 3.16 starts with the method `checkString()` that checks whether or not the string `sub_str` is a substring of `my_str` whose lengths are assigned to the integer variables sub_len and str_len, respectively. The next portion of Listing 3.16 consists of a loop that iterates through the characters in the string substr whose index position is at most str_len-sub_len-pos so that it does not exceed the length of substr. Notice that the loop uses a "sliding window" to compare a substring of my_str whose length equals the length of sub_str: if they are equal than a repeated string exists in my_str, and the results are returned.

The next portion of Listing 3.16 defines the `main()` method that starts by initiating the string array my_strings with a set of strings, each of which will be examined to determine whether or not it contains a repeated substring of length at least 2. The next portion of the `main()` method consists of an outer loop that iterates through each element of the my_strings array.

During each iteration, a nested loop processes the current element containing a nested loop that iterates from index position 0 up to the midpoint of the current element. Next, a substring of the current element is extracted and processed by the check_string() method to determine whether or not there is a matching substring. In addition, the returned values are compared with the current "candidate" for the maximum repeated substring, and the values are updated accordingly.

A final conditional code block displays a message depending on whether or not the current element in my_strings contain a maximum repeated substring. Now launch the code in Listing 3.16 and you will see the following output:

```
==> Repeating substrings have length >= 2
=> Checking current string:     abc
No maximum repeating substring: abc

=> Checking current string:     abab
=> Maximum repeating substring: ab

=> Checking current string:     abccba
No maximum repeating substring: abccba

=> Checking current string:     azaaza
=> Maximum repeating substring: aza

=> Checking current string:     abcdefgabccbaxyz
=> Maximum repeating substring: abc
```

This concludes the portion of the chapter regarding strings-related code samples. The next section introduces you to 1D arrays and related code samples.

WORKING WITH 1D ARRAYS

A *one-dimensional array* in `Java` is a one-dimensional construct whose elements are homogeneous (i.e., mixed data types are not permitted). Given two arrays A and B, you can add or subtract them, provided that they have the same number of elements. You can also compute the inner product of two vectors by calculating the sum of their component-wise products.

Now that you understand some of the rudimentary operations with one-dimensional matrices, the following subsections illustrate how to perform various tasks on arrays in `Java`.

Rotate an Array

Listing 3.17 displays the contents of `RotateArray.java` that illustrates how to rotate the elements in an array.

LISTING 3.17: RotateArray.java

```
import java.util.ArrayList;

public class RotateArray
{
   public static void main(String[] args)
   {
      int temp=0, item=0, saved=0;
      int[] arr1 = new int[]{5,10,17,23,30,47,50};
      int arr1_len = arr1.length;
      System.out.println("arr1_len: "+arr1_len);

      System.out.println("original:");
      for(int j=0; j<arr1.length; j++) {
         System.out.print(arr1[j]+" ");
      }

      System.out.println();
      int shift_count = 2;
      for(int i=0; i<shift_count; i++) {
        saved = arr1[0];
        for(int j=1; j<arr1.length; j++) {
          arr1[j-1] = arr1[j];
        }
        arr1[arr1.length-1] = saved;
      }

      System.out.println("rotated:");
      for(int j=0; j<arr1.length; j++) {
         System.out.print(arr1[j]+" ");
      }
      System.out.println();
   }
```

Listing 3.17 initializes the variable `list` as a list of integers and the variable `shift_count` with the value 2: the latter is the number of positions to shift leftward the elements in `list`. The next portion of Listing 3.17 is a loop that performs two actions:

1. It "pops" the left-most element of the list.
2. It appends that element to the list so that it becomes the new right-most element.

The loop is executed `shift_count` iterations, after which the elements in list have been rotated the specified number of times. Launch the code in Listing 3.18 and you will see the following output:

```
original: [5, 10, 17, 23, 30, 47, 50]
rotated:  [17, 23, 30, 47, 50, 5, 10]
```

TASK: INVERT ADJACENT ARRAY ELEMENTS

Listing 3.18 displays the contents of `InvertItems.java` that illustrates how to perform a pair-wise inversion of adjacent elements in an array. Specifically, index 0 and index 1 are switched, then index 2 and index 3 are switched, and so forth until the end of the array.

LISTING 3.18: InvertItems.java

```
import java.util.Arrays;

public class InvertItems
{
   public static void main(String[] args)
   {
      int temp=0, mid_point=0;
      int[] arr1 = new int[]{5,10,17,23,30,47,50};

      System.out.println("Original: "+Arrays.
      toString(arr1));

      mid_point = (int)(arr1.length/2);

      for(int idx=0; idx<mid_point+2; idx+=2)
      {
        temp = arr1[idx];
        arr1[idx] = arr1[idx+1];
        arr1[idx+1] = temp;
      }

      System.out.println("Inverted: "+Arrays.
      toString(arr1));
   }
}
```

Listing 3.18 starts with the array `arr1` of integers and the variable `mid_point` that is the mid-point of `arr1`. The next portion of Listing 3.18 contains a loop that iterates from index 0 to index `mid_point+2`, where the loop variable `ndx` is incremented by 2 (not by 1) after each iteration. As you can see, the code performs a standard "swap" of the contents of `arr1` in index positions `ndx` and `ndx+1` by means of a temporary variable called `temp`. Launch the code in Listing 3.18 and you will see the following output:

```
original: [ 5 10 17 23 30 47 50]
inverted: [10  5 23 17 47 30 50]
```

Listing 3.19 displays the contents of `Swap.java` that illustrates how to invert adjacent values in an array without using an intermediate temporary variable. Notice that Listing 3.18 does not require a temporary variable `temp` to switch adjacent values, whereas Listing 3.19 (shown below) depends on a temporary intermediate variable.

LISTING 3.19: Swap.java

```
import java.util.Arrays;

public class Swap
{
   public static int[] swap(int num1, int num2)
   {
     int delta;
     delta = num2 - num1;
     //print("num1:",num1,"num2:",num2)

     num2 = delta;
     num1 = num1+delta;
     num2 = num1-delta;
     return new int[]{num1,num2};
   }

   public static void main(String[] args)
   {
      int num1, num2;
      int[] result;

      int[] arr1 = new int[] {15,4,23,35,80,50};
      System.out.println("BEFORE: "+Arrays.toString(arr1));


      for(int idx=0; idx<arr1.length; idx+=2)
      {
        result = swap(arr1[idx],arr1[idx+1]);
        num1 = result[0];
        num2 = result[1];
        arr1[idx]   = num1;
        arr1[idx+1] = num2;
      }

      System.out.println("AFTER:  "+Arrays.toString(arr1));
   }
}
```

Listing 3.19 starts with the method `swap()` that switches the values of two numbers. If this section of code is not clear, try manually executing this code with hard-coded values for `num1` and `num2`. The next portion of Listing 3.19 initializes the array `arr1` with integers, followed by a loop that iterates through the values of `arr1` and invoking the method `swap()` during each iteration. Launch the code in Listing 3.19 and you will see the following output:

```
BEFORE arr1: [15  4 23 35 80 50]
AFTER  arr1: [ 4 15 35 23 50 80]
```

TASK: SHIFT NONZERO ELEMENTS LEFTWARD

Listing 3.20 displays the contents of `ShiftZeroesLeft.java` that illustrates how to shift nonzero values toward the left while maintaining their relative positions.

LISTING 3.20: ShiftZeroesLeft.java

```
public class ShiftZeroesLeft
{
   public static void main(String[] args)
   {
      int left=-1;
      int[] arr1 = new int[]{0,10,0,0,60,30,0,200,0};

      System.out.println("Initial:");
      for(int i=0;i<arr1.length; i++) {
         System.out.print(arr1[i]+" ");
      }
      System.out.println("\n");

      // find the left-most index with value 0:
      for(int i=0;i<arr1.length; i++) {
         if(arr1[i] == 0) {
           left = i;
         } else {
           left += 1;
           break;
         }
      }
      System.out.println("non-zero index: "+left);

      // ex: 0 10 0 0 30 60 0 200 0
      // right shift positions left-through-(idx-1):
      for(int idx=left+1; idx<arr1.length; idx++) {
         if(arr1[idx] == 0) {
           for(int j=idx-1;j>=left; j--) {
             arr1[j+1] = arr1[j];
           }
           arr1[left] = 0;
          System.out.println("shifted non-zero position "+left);
           left += 1;
         }
      }

      System.out.println("\nSwitched:");
      for(int i=0;i<arr1.length; i++) {
```

```
            System.out.print(arr1[i]+" ");
         }
         System.out.println();
      }
   }
```

Listing 3.20 starts with a `main()` function that initializes the array arr1 with a set of ten integers, followed by a loop that displays the contents of arr1. The second loop iterates through the values of arr1 in order to determine the index of the left-most nonzero value in arr1, which is assigned to the variable left.

The third loop starts from index left+1 and performs a right shift of the nonzero values in arr1 that have a smaller index position (i.e., on the left side). Next, the value of arr1[left] is initialized with 0, thereby completing the right-shift of nonzero values. This process continues until the right-most element of arr1 has been processed. Launch the code in Listing 3.20 and you will see the following output:

```
Initial:
0 10 0 0 30 60 0 200 0
non-zero index:  1
shifted non-zero position  1
shifted non-zero position  2
shifted non-zero position  3
shifted non-zero position  4
switched:
0 0 0 0 0 30 30 60 200
```

TASK: SORT ARRAY IN-PLACE IN O(N) WITHOUT A SORT FUNCTION

Listing 3.21 displays the contents of `SimpleSort.java` that illustrates a very simple way to sort an array containing an ***equal*** number of values 0, 1, and 2 without using another data structure.

LISTING 3.21: SimpleSort.java

```
public class SimpleSortColors
{
   public static void main(String[] args)
   {
      int[] arr1 = new int[]{0,1,2,2,1,0,0,1,2};
      int zeroes = 0;

      System.out.println("Initial:");
      for(int i=0;i<arr1.length; i++) {
         System.out.print(arr1[i]+" ");
      }
      System.out.println();
```

```
        int third=arr1.length/3;
        for(int i=0;i<third; i++) {
           arr1[i]          = 0;
           arr1[third+i]    = 1;
           arr1[2*third+i]  = 2;
        }

        System.out.println("Sorted:");
        for(int i=0;i<arr1.length; i++) {
           System.out.print(arr1[i]+" ");
        }
        System.out.println();
    }
}
```

Listing 3.21 contains a `main()` method that initializes the array `arr1` with an equal number of entries for 0, 1, and 2. Therefore, `arr1` must have length `3*n`, where `n` is a positive integer. Therefore, we use a loop that sets the first third of `arr1` with 0, the second third with the value 1, and the final third with the value 2.

We can execute the preceding loop because the key word in this task description is *equal*, and therefore there is no need for another data structure, or temporary variables, or any comparisons between pairs of elements in the array arr1. Now launch the code in Listing 3.21 and you will see the following output:

```
Initial:
0 1 2 2 1 0 0 1 2
Sorted:
0 0 0 1 1 1 2 2 2
```

TASK: GENERATE 0 THAT IS THREE TIMES MORE LIKELY THAN A 1

The solution to this task is based on the observation that an AND operator with two inputs generates three 0s and a single 1: we only need to randomly generate a 0 and a 1 and supply those random values as input to the AND operator.

Listing 3.22 displays the contents of `GenerateZeroOrOne.java` that illustrates how to generate 0s and 1s with the expected frequency.

LISTING 3.22: GenerateZeroOrOne.java

```
public class GenerateZeroOrOne
{
   public static void main(String[] args)
   {
      int zeroes=0, ones=0, max=100;

      for(int i=0; i<max; i++) {
         int x  = (int)Math.floor(Math.random()*2);
```

```
         int y  = (int)Math.floor(Math.random()*2);
         int xy = (x & y);

         if(xy == 0) ones++;
         if(xy == 1) zeroes++;
      }

      System.out.println("Percentages ("+max+" iterations):");
      System.out.println("ones:  "+(100*ones/max));
      System.out.println("zeoes: "+(100*zeroes/max));
   }
}
```

Listing 3.22 defines a `main()` method that initializes the integer-valued variables zeroes, ones, and max with the values 0, 0, and 100, respectively. The next portion of code contains a loop that iterates from 0 to max. Each iteration initializes the variables x and y with a random integer value that is either 0 or 1, and then initializes the variable xy with the logical OR of x and y. The next portion of the loop increments either ones or zeroes depending on whether xy equals 0 or 1, respectively. The last code block in Listing 3.22 displays the values of max, ones, and zeros. Launch the code in Listing 3.22 and you will see the following output:

```
Percentages (100 iterations):
ones:  74
zeoes: 26
```

TASK: INVERT BITS IN EVEN AND ODD POSITIONS

This solution to this task involves two parts: the first part "extracts" the bits in the even positions and shifts the result one bit to the right, followed by the second part that extracts the bits in the odd positions and shifts the result one bit to the right. Listing 3.23 displays the contents of `SwitchBitPairs.java` that illustrates how to solve this task.

LISTING 3.23: SwitchBitPairs.java

```
public class SwitchBitPairs
{
   public static void main(String[] args)
   {
     // This task involves two binary masks:
     // A for even bits and 5 for odd bits
     // hex A = 8 + 2 = 1010
     // hex 5 = 4 + 1 = 0101
     // 01010101 1+4+16+64 = 85
     // 10101010 2+8+32+128 = 170

     int[] numbers = new int[]{42, 37, 52, 63};
```

```
      for(int num : numbers) {
         String bnum =  Integer.toBinaryString(num);
         int rnum = ((num & 0xaa) >> 1) | ((num & 0x55) << 1);
         String srnum = Integer.toBinaryString(rnum);

         String srnum2 = String.format("%6s",
                 Integer.toBinaryString(rnum)).replace(' ','0');

         System.out.println("num:   "+num+ " bnum:   "+bnum);
         System.out.println("rnum: "+rnum+
                        " srnum: "+srnum+" srnum2: "+srnum2);
         System.out.println("----------------\n");
      }
   }
}
```

Listing 3.23 defines a `main()` method that initializes the integer-valued variable numbers with an array of four integers. The next portion of code contains a loop that iterates through the elements of the array numbers. Each iteration initializes the variables `rnum` as the binary representation of `num` (the loop variable). Next, the integer-valued variable `num` is initialized as the logical OR of two quantities, as shown here:

```
int rnum = ((num & 0xaa) >> 1) | ((num & 0x55) << 1);
```

The left term in the preceding code snippet performs a logical AND of num and the hexadecimal number 0xaa, and then performs a right shift. The right term in the preceding code snippet performs a logical AND of `num` and the hexadecimal number 0x55, and then performs a left shift.

The next code snippet in Listing 3.23 initializes the variable sum with the string-based binary representation of `rnum`, followed by a code snippet that replaces blank spaces in `rnum` with "0". The final portion of Listing 3.24 displays the values of all the variables that are initialized in the loop. Now launch the code in Listing 3.23 and you will see the following output:

```
num:  42 bnum:  101010
rnum: 21 srnum: 10101 srnum2: 010101
----------------


num:  37 bnum:  100101
rnum: 26 srnum: 11010 srnum2: 011010
----------------


num:  52 bnum:  110100
rnum: 56 srnum: 111000 srnum2: 111000
----------------


num:  63 bnum:  111111
rnum: 63 srnum: 111111 srnum2: 111111
----------------
```

TASK: CHECK FOR ADJACENT SET BITS IN A BINARY NUMBER

This solution to this task involves Listing 3.24 displays the contents of the `Java` file `CheckAdjacentBits.java` that illustrates how to solve this task.

LISTING 3.24: CheckAdjacentBits.java

```
// true if adjacent bits are set in num:

public class CheckAdjacentBits
{
   // true if adjacent bits are set in bin(num):
   public static boolean check(int num) {
      return (num & (num << 1)) != 0;
   }

   public static void main(String[] args)
   {
      int[] arr = new int[]{9,18,31,67,88,100};

      for(int num : arr) {
         System.out.println(num+" (binary): "+
                                 Integer.toBinaryString(num));

         if (check(num)) {
           System.out.println(num+": found adjacent set bits\n");
         }
         else {
            System.out.println(num+": no pair of adjacent
            set bits\n");
         }
      }
   }
}
```

Listing 3.24 defines the static method `check()` with a single line of code that returns a Boolean value, which determines whether or not the logical AND of the integer num with the right-shifted result of num is nonzero.

The next portion of Listing 3.24 defines a `main()` method that initializes the integer-valued variable arr with an array of six integers. The next portion of code contains a loop that iterates through the elements of the array arr. Each iteration tests the result of invoking the method `check()` with the variable num (the loop variable). A suitable message is printed depending on whether or not the return value is True or False. Launch the code in Listing 3.24 and you will see the following output:

```
9 (binary): 1001
9: no pair of adjacent set bits
```

```
18 (binary): 10010
18: no pair of adjacent set bits

31 (binary): 11111
31: found adjacent set bits

67 (binary): 1000011
67: found adjacent set bits

88 (binary): 1011000
88: found adjacent set bits

100 (binary): 1100100
100: found adjacent set bits
```

TASK: COUNT BITS IN A RANGE OF NUMBERS

This solution to this task involves Listing 3.25 displaying the contents of the Java file `BitCount.java` that illustrates how to solve this task.

LISTING 3.25: BitCount.java

```
public class BitCount
{
   public static void main(String[] args)
   {
      int bits = 0;
      int[] numbers = new int[]{1,7,8,15,16,17,33,64,256};

      for(int num : numbers) {
         bits = (int)Math.ceil((Math.log(num+1)/ Math.
         log(2)));
         if((num == 1) && (bits == 0)) bits = 1;
         System.out.println("Number: "+num+" bit count:
         "+bits);
      }
   }
}
```

Listing 3.25 defines a `main()` method that initializes the integer-valued variable bits as 0 and the integer-valued array numbers with an array of nine integers. The next portion of code contains a loop that iterates through the elements of the array numbers. Each iteration initializes the variable bits with the integer-valued portion of ceiling of the ratio of two logarithmic values, as shown here:

```
bits = (int)Math.ceil((Math.log(num+1)/ Math.log(2)));
```

The next code snippet sets bits equal to 1 if num equals 1 and bits are equal to 0 (this is a "corner case" for the loop). The final code snippet is a print statement that displays the values of num and bits. Launch the code in Listing 3.25 and you will see the following output:

```
Number: 1 bit count: 1
Number: 7 bit count: 3
Number: 8 bit count: 4
Number: 15 bit count: 4
Number: 16 bit count: 5
Number: 17 bit count: 5
Number: 33 bit count: 6
Number: 64 bit count: 7
Number: 256 bit count: 9
```

TASK: FIND THE RIGHT-MOST SET BIT IN A NUMBER

This solution to this task involves Listing 3.26 displays the contents of `RightMostSetBit.java` that illustrates how to solve this task.

LISTING 3.26: RightMostSetBit.java

```
public class RightMostSetBit
{
   public static void main(String[] args)
   {
      int[] numbers = new int[]{42, 37, 52, 63};

      for(int num : numbers) {
         int two1 =  ~(num-1);
         int two2 =  (num & ~(num-1));

         String padn = String.format("%6s",
                Integer.toBinaryString(num)).replace(' ', '0');

         System.out.println(
             "num: "+num+" two1: "+two1+" two2: "+two2+"
             padn: "+padn);
      }

      System.out.println();
   }
}
```

Listing 3.26 defines a `main()` method that initializes the integer-valued array numbers with an array of four integers. The next portion of code contains a loop that iterates through the elements of the array numbers. Each iteration initializes the variables two1 and two2 based on a bit-based operation involving the complement of (num-1) and the logical and of num and the preceding quantity for two1 and two2, respectively.

The nest portion of listing 3.26 initializes the string-valued variable padn that replaces space characters in num with the string "0." The final code snippet prints the values of num, two, and two. Now launch the code in Listing 3.26 and you will see the following output:

```
num: 42 two1: -42 two2: 2 padn: 101010
num: 37 two1: -37 two2: 1 padn: 100101
num: 52 two1: -52 two2: 4 padn: 110100
num: 63 two1: -63 two2: 1 padn: 111111
```

TASK: THE NUMBER OF OPERATIONS TO MAKE ALL CHARACTERS EQUAL

This solution to this task involves Listing 3.27 displays the contents of the `Java` file `FlipBitCount.java` that illustrates how to solve this task.

LISTING 3.27: FlipBitCount.java

```
public class FlipBitCount
{
   // determine the minimum number of operations
   // to make all characters of the string equal
   public static int minOperations(String the_string)
   {
      int count = 0; // track the # of changes

      for(int i=1; i<the_string.length(); i++) {
        // are adjacent characters equal?
        if (the_string.charAt(i) != the_string.charAt(i-1))
          count += 1;
      }

      return(count);
   }

   public static void main(String[] args)
   {
      String[] arr1 = new String[]
          {"0101010101", "1111010101", "100001", "111111"};

      for(String str1 : arr1)
      {
         System.out.println("String: "+str1);
         System.out.println("Result: "+ minOperations(str1));
         System.out.println("----------------\n");
      }
   }
}
```

Listing 3.27 defines the static method `cminOperations()` that counts the number of adjacent (contiguous) characters that are different via a loop that iterates through the characters in the string the_string and performs a simple comparison of adjacent characters.

The next portion of Listing 3.27 initializes the string-based variable arr1 and then defines a `main()` method that iterates through the elements of the array numbers. Each iteration displays the current string and the result of invoking the minOperations() method with the current string. Launch the code in Listing 3.27 and you will see the following output:

```
String:  0101010101
Result:  9
-----------------

String:  1111010101
Result:  6
-----------------

String:  100001
Result:  2
-----------------

String:  111111
Result:  0
-----------------
```

TASK: COMPUTE XOR WITHOUT XOR FOR TWO BINARY NUMBERS

This solution to this task involves Listing 3.28 displays the contents of the Java file `XORWithoutXOR.java` that illustrates how to solve this task.

LISTING 3.28: XORWithoutXOR.java

```
class XORWithoutXOR
{
   public static int findBits(int x, int y)
   {
     return (x | y) - (x & y);
   }

   public static void main(String[] args)
   {
      int[] arrx = new int[]{65,15};
      int[] arry = new int[]{80,240};

      for(int i=0; i<arrx.length; i++)
      {
         int x = arrx[i];
         int y = arry[i];

         String xstr  = Integer.toBinaryString(x);
         String ystr  = Integer.toBinaryString(y);
```

```
            String xory   = Integer.toBinaryString(x|y);
            String xandy  = Integer.toBinaryString(x&y);
            String xxory  = Integer.toBinaryString(findBits(x,y));

            System.out.println("Decimal x: "+x);
            System.out.println("Decimal y: "+y);

            System.out.println("Binary x:  "+xstr);
            System.out.println("Binary y:  "+ystr);

            System.out.println("x OR y:    "+xory);
            System.out.println("x AND y:   "+xandy);
            System.out.println("x XOR y:   "+xxory);
            System.out.println("-----------------\n");
        }
    }
}
```

Listing 3.28 starts with the static method findBits() that returns the XOR of two numbers with the following code snippet that does not require an XOR operation:

```
return (x | y) - (x & y);
```

The next portion of Listing 3.28 defines a `main()` method that initializes two integer arrays arrx and arry, followed by a loop that processes the elements of `arrx` and `arry` in a pair-wise fashion. During each iteration, the variables x and y are initialized to the current pair of values in `arrx` and `arry`, followed by the variables xstr and ystr that are initialized as the string-based counterparts of the variables x and y.

The next portion of the loop initializes the variables `xory` and `xandy` as the string-based values for the OR and AND of the variables x and y, respectively. In addition, the variable and `xxory` is initialized by the result of invoking the findBits() method with x and y, which is simply the XOR of x and y. The next portion of the loop displays the decimal and binary values of x and y, followed by the values of `xory`, `xandy`, and `xxory`. Launch the code in Listing 3.28 and you will see the following output:

```
Decimal x: 65
Decimal y: 80
Binary x:  1000001
Binary y:  1010000
x OR y:    1010001
x AND y:   1000000
x XOR y:   10001
-----------------

Decimal x: 15
Decimal y: 240
Binary x:  1111
```

```
Binary y:  11110000
x OR y:    11111111
x AND y:   0
x XOR y:   11111111
------------------
```

TASK: SWAP ADJACENT BITS IN TWO BINARY NUMBERS

Listing 3.29 displays the contents of `SwapAdjacentBits.java` that illustrates how to solve this task.

LISTING 3.29: SwapAdjacentBits.java

```
public class SwapAdjacentBits
{
   public static String toBinaryString(int n)
   {
      return String.format("%32s", Integer.toBinaryString(n))
                  .replaceAll(" ", "0");
   }

   // Function to swap adjacent bits of a given number
   public static int swapAdjacentBits(int n) {
       return (((n & 0xAAAAAAAA) >>1)|((n & 0x55555555) <<1));
   }

   public static void main(String[] args)
   {
      int[] numbers = new int[]{17, 761622921, 123};

      for (int num : numbers)
      {
        System.out.println(num + " binary: " +
                           toBinaryString(num));

        int num2 = swapAdjacentBits(num);
        System.out.println("After Swapping:");
        System.out.println(num + " binary: " +
                           toBinaryString(num2)+"\n");
      }
   }
}
```

Listing 3.29 starts with the definition of the function `toBinaryString()` generates a binary 32-bit string that is left-padded with 0. The next portion of Listing 3.29 defines the function `swapAdjacentBits()` that swaps adjacent bits by performing an OR operation on two quantities. The left quantity calculates the AND of an integer n with the value `0xAAAAAAAA` and then shifts the result one bit to the right. The right quantity calculates the AND of an integer n with the value 0x55555555 and then shifts the result one bit to the left. The result is returned by the swapAdjacentBits() method, as shown here:

```
return (((n & 0xAAAAAAAA) >>1)|((n & 0x55555555) <<1));
```

The next portion of Listing 3.29 defines the main() method that initializes the array numbers with a set of numbers, followed by a loop that iterates through the values in numbers. During each iteration, the binary value of the current number number is displayed, followed by the binary value of the "flipped" number. Now launch the code in Listing 3.29 and you will see the following output:

```
17 binary: 00000000000000000000000000010001
After Swapping:
17 binary: 00000000000000000000000000100010

761622921 binary: 00101101011001010111000110001001
After Swapping:
761622921 binary: 00011110100110101011001001000110

123 binary: 00000000000000000000000001111011
After Swapping:
123 binary: 00000000000000000000000010110111
```

WORKING WITH 2D ARRAYS

A *two-dimensional array* in Java is a two-dimensional construct whose elements are homogeneous (i.e., mixed data types are not permitted). Given two arrays A and B, you can add or subtract them, provided that they have the same number of rows and columns.

Multiplication works differently: if A is an mxn matrix that you want to multiply (on the right of A) by B, then B must be a nxp matrix. The rule for matrix multiplication is as follows: the number of columns of A must equal the number of rows of B.

In addition, the *transpose* of matrix A is another matrix At such that the rows and columns are interchanged. Thus, if A is an mxn matrix then At is an nxm matrix. The matrix A is symmetric if A = At. The matrix A is the identity matrix I if the values in the main diagonal (upper left to lower right) are 1 and the other values are 0. The matrix A is invertible if there is a matrix B such that A*B = B*A = I. Based on the earlier discussion regarding the product of two matrices, both A and B must be square matrices with the same number of rows and columns.

Now that you understand some of the rudimentary operations with strings, the following subsection illustrates how to perform various tasks on matrices in Java.

THE TRANSPOSE OF A MATRIX

As a reminder, the transpose of matrix `A` is matrix `At`, where the rows and columns of `A` are the columns and rows, respectively, of matrix `At`.

Listing 3.30 displays the contents of `MatTranspose.java` that illustrates how to find the transpose of an `mxn` matrix.

LISTING 3.30: MatTranspose.java

```
public class MatTranspose
{
   public static int[][] transpose(int[][]A, int rows, int cols)
   {
      for(int i=0; i<rows; i++) {
        for(int j=i; j<cols; j++) {
          //System.out.println("switching "+A[i][j]+" and
          "+A[j][i]);
          int temp = A[i][j];
          A[i][j] = A[j][i];
          A[j][i] = temp;
        }
      }
      return A;
   }

   public static void main(String[] args)
   {
      int[][] A1 = new int[][]{{100,3},{500,7}};
      System.out.println("=> original:");
      System.out.println(A1);

      int[][] At1 = transpose(A1, 2, 2);
      System.out.println("=> transpose:");
      System.out.println(At1);
      System.out.println();

      // example 2:
      int[][] A2 = new int[][]{{100,3,-
      1},{30,500,7},{123,456,789}};
      System.out.println("=> original:");
      System.out.println(A2);

      int[][] At2 = transpose(A2, 3, 3);
      System.out.println("=> transpose:");
      System.out.println(At2);
   }
}
```

Listing 3.30 is actually straightforward: the function `transpose()` contains a nested loop that uses a temporary variable `temp` to perform a simple swap of the values of `A[i,j]` and `A[j,i]` in order to generate the transpose of the matrix `A`. The next portion of Listing 3.30 initializes a 2x2 array `A` and then invokes the function `transpose` to generate its transpose. Launch the code in Listing 3.30 and you will see the following output:

```
=> original:
[[100   3]
 [500   7]]
=> transpose:
[[100 500]
 [  3   7]]

=> original:
[[100   3  -1]
 [ 30 500   7]
 [123 456 789]]
=> transpose:
[[100  30 123]
 [  3 500 456]
 [ -1   7 789]]
```

In case you didn't notice, the transpose `At` of a matrix `A` is actually a 90 degree rotation of matrix `A`. Hence, if `A` is a square matrix of pixels values for a `PNG`, then `At` is a 90 degree rotation of the `PNG`. However, if you take the transpose of `At`, the result is the original matrix `A`.

SUMMARY

This chapter started with an introduction to one-dimensional vectors, and how to calculate their length and the inner product of pairs of vectors. Then you saw how to perform various tasks involving numbers, such as multiplying and dividing two positive integers via recursive addition and subtraction, respectively.

In addition, you learned about working with strings, and how to check a string for unique characters, how to insert characters in a string, and how to find permutations of a string. Next, you learned about determining whether or not a string is a palindrome.

Moreover, you learned how to calculate the transpose of a matrix, which is the equivalent of rotating a bitmap of an image by 90 degrees.

CHAPTER 4

Search and Sort Algorithms

The first half of this chapter provides an introduction to some well-known search algorithms, followed by the second half that discusses various sorting algorithms.

The first section of this chapter introduces search algorithms such as linear search and binary search, that you can use when searching for an item (which can be numeric or character-based) in an array. A linear search is inefficient because it requires an average of `n/2` (which has complexity O[n]) comparisons to determine whether or not the search element is in the array, where n is the number of elements in the list or array.

By contrast, a binary search required O (log n) comparisons, which is vastly more efficient with larger sets of items. For example, if an array contains 1,024 items, a most ten comparisons are required in order to find an element because 2**10 = 1024, so log(1024) = 10 (using base 2). However, a binary search algorithm requires a sorted list of items.

The second part of this chapter discusses some well-known sorting algorithms, such as the bubble sort, selection sort, insertion sort, the merge sort, and the quicksort that you can perform on an array of items.

SEARCH ALGORITHMS

The following list contains some well-known search algorithms that will be discussed in several subsections:

- linear search
- binary search
- jump search
- Fibonacci search

A *linear search* algorithm is probably the simplest of all the search algorithms: this algorithm checks every element in an array until either the desired item is located or the end of the array is reached.

However, as you learned in the introduction for this chapter, a linear search is inefficient when an array contains a large number of values. If the array is very small, the difference in performance between a linear search and a binary search can also be very small; in this case, a linear search might be an acceptable choice of algorithms.

In the RDBMS (relational database management system) world, tables often have an index (and sometimes more than one) in order to perform a table search more efficiently. However, there is some additional computational overhead involving the index, which is a separate data structure that is stored on disk. However, a linear search involves only the data in the table. As a rule of thumb, an index-based search is more efficient when tables have more than 300 rows (but results can vary). The next section contains a code sample that performs a linear search on an array of numbers.

Linear Search

Listing 4.1 displays the contents of the `LinearSearch.java` that illustrates how to perform a linear search on an array of numbers.

LISTING 4.1: LinearSearch.java

```
public class LinearSearch
{
   public static void main(String[] args)
   {
      int found = -1, item = 123;
      int[] arr1 = new int[]{1,3,5,123,400};

      for(int i=0; i<arr1.length; i++) {
        if (item == arr1[i]) {
           found = i;
           break;
        }
      }

      if (found >= 0)
        System.out.println("found "+item+" in position
        "+found);
      else
        System.out.println(item+" not found");
   }
}
```

Listing 4.1 starts with the variable `found` that is initialized with the value -1, followed by the search item 123, and also the array `arr1` that contains an array of numbers. Next, a loop that iterates through the elements of the array `arr1` of integers, comparing each element with the value of item. If a match occurs, the variable `found` is set equal to the value of the loop variable `i`, followed by an early exit.

The last portion of Listing 4.1 checks the value of the variable `found`: if it's non-negative then the search item was found (otherwise it's not in the array). Launch the code in Listing 4.1 and you will see the following output:

```
found 123 in position 3
```

Keep in mind the following point: although the array `arr1` contains a sorted list of numbers, the code works correctly for an unordered list as well.

Binary Search Walk-Through

A *binary search* requires a sorted array and can be implemented via an iterative algorithm as well as a recursive solution. The key idea involves comparing the middle element of an array of sorted elements with a search element. If they are equal, then the item has been found; if the middle element is smaller than the search element, repeat the previous step with the right half of the array; if the middle element is larger than the search element, repeat the previous step with the left half of the array. Eventually the element will be found (if it appears in the array) or the repeated splitting of the array terminates when the subarray has a single element (i.e., no further splitting can be performed).

Let's perform a walk-through of a binary search that searches for an item in a sorted array of integers.

Example #1: let `item = 25` and `arr1 = [10,20,25,40,100]`, so the midpoint of the array is 3. Since `arr1[3] == item`, the algorithm terminates successfully.

Example #2: let `item = 25` and `arr1 = [1,5,10, 15, 20, 25, 40]`, which means that the midpoint is 4.
First iteration: since arr1[4] < item, we search the array [20,25,40]
Second iteration: the midpoint is 1, and the corresponding value is 25.
Third iteration: 25 and the array is the single element [25], which matches the item.

Example #3: item = 25 and `arr1 = [10, 20, 25, 40, 100,150,400]`, so the midpoint is 4.
First iteration: since arr1[4] > 25, we search the array [10,20,25].
Second iteration: the midpoint is 1, and the corresponding value is 20.
Third iteration: 25 and the array is the single element [25], which matches the item.

Example #4: item = 25 and `arr1 = [1,5,10, 15, 20, 30, 40]`, so the midpoint is 4.
First iteration: since arr1[4] < 25, we search the array [20,30,40].
Second iteration: the midpoint is 1, and the corresponding value is 30.
Third iteration: 25 and the array is the single element [20], so there is no match.

As mentioned in the first paragraph of this section, a binary search can be implemented with an interactive solution, which is the topic of the next section.

Binary Search (Iterative Solution)

Listing 4.2 displays the contents of `BinarySearch.java` that illustrates how to perform a binary search with an array of numbers.

LISTING 4.2: BinarySearch.java

```
public class BinarySearch
{
   public static void main(String[] args)
   {
      int[] arr1 = new int[]{1,3,5,123,400};
      int left = 0, found = -1, item = 123;
      int right = arr1.length-1;

      System.out.print("array: ");
      for(int i=0; i<arr1.length; i++) {
        System.out.print(arr1[i]+" ");
      }
      System.out.println("");

      while(left <= right) {
        int mid = (int)(left + (right-left)/2);

        if(arr1[mid] == item) {
          found = mid;
          break;
        } else if (arr1[mid] < item) {
          left = mid+1;
        } else {
          right = mid-1;
        }
      }

      if( found >= 0)
        System.out.println("found "+item+" in position
        "+found);
      else
       System.out.println(item+" not found");
   }
}
```

Listing 4.2 initializes an array of numbers and some scalar variables to keep track of the left and right index positions of the subarray that we will search each time that we split the array. The next portion of Listing 4.2 contains conditional logic that implements the sequence of steps that you saw in the examples in the previous section. Launch the code in Listing 4.2 and you will see the following output:

```
array: [   1    3    5 123 400]
found 123 in position 3
```

Binary Search (Recursive Solution)

Listing 4.3 displays the contents of `BinarySearchRecursive.java` that illustrates how to perform a binary search recursively with an array of numbers.

LISTING 4.3: BinarySearchRecursive.java

```
public class BinarySearchRecursive
{
   public static Boolean binarySearch(int[] data, int item,
   int left, int right)
   {
     if( left > right ) {
       return false;
     } else {
       //incorrect (can result in overflow):
       //int mid = (left + right) / 2
       int mid = (int)(left + (right-left)/2);

       if(item == data[mid]) {
         return true;
       } else if(item < data[mid]) {
         //recursively search the left half
         return binarySearch(data, item, left, mid-1);
       } else {
         //recursively search the right half
         return binarySearch(data, item, mid+1, right);
       }
     }
   }

   public static void main(String[] args)
   {
      int[] items = new int[]{-100, 123, 200, 400};
      int[] arr1 = new int[]{1,3,5,123,400};
      int[] arr2 = new int[]{1,3,5,123,400};
      int found = -1, item = 123;
      int left = 0, right = arr1.length-1;

      System.out.print("array: ");
      for(int i=0; i<arr1.length; i++) {
        System.out.print(arr1[i]+" ");
      }
      System.out.println("");
```

```
        for(Integer item2 : items) {
          arr1 = arr2;
          System.out.println("searching for item:  "+item2);
          left  = 0;
          right = arr1.length-1;
          Boolean result = binarySearch(arr1, item2, left,
          right);
         System.out.println("item:  "+item2+ " found: "+result);
        }

        if( found >= 0)
          System.out.println("found "+item+" in position
          "+found);
        else
         System.out.println(item+" not found");
    }
}
```

Listing 4.3 starts with the method `binarySearch()` with parameters `data, item, left`, and `right)` that contain the current array, the search item, the left index of `data`, and the right index of `data`, respectively. If the left index `left` is greater than the right index `right` then the search item does not exist in the original array.

However, if the left index `left` is *less than* the right index `right` then the code assigns the middle index of `data` to the variable `mid`. Next, the code performs the following three-part conditional test:

If `item == data[mid]` then the search item has been found in the array.
If `item < data[mid]` then the function `binary_search()` is invoked with the left-half of the `data` array.
If `item > data[mid]` then the function `binary_search()` is invoked with the right-half of the `data` array.

The next portion of Listing 4.3 initializes the sorted array `arr1` with numbers and initializes the array `items` with a list of search items, and also initializes some scalar variables to keep track of the left and right index positions of the subarray that we will search each time that we split the array.

The final portion of Listing 4.3 consists of a loop that iterates through each element of the items array and invokes the method `binarySearch()` to determine whether or not the current item is in the sorted array. Launch the code in Listing 4.3 and you will see the following output:

```
array: 1 3 5 123 400
searching for item:  -100
item:  -100 found: false
searching for item:  123
item:  123 found: true
searching for item:  200
item:  200 found: false
```

```
searching for item:  400
item:  400 found: true
123 not found
```

WELL-KNOWN SORTING ALGORITHMS

Sorting algorithms have a best case, average case, and worst case in terms of performance. Interestingly, sometimes an efficient algorithm (such as quicksort) can perform the worst when a given array is already sorted.

The following subsections contain code samples for the following well-known sort algorithms:

- bubble sort
- selection sort
- insertion sort
- merge sort
- quicksort
- bucket sort
- Shell sort
- Shell sort 108
- heap sort 108
- bucket sort
- in-place sort
- counting sort
- radix sort

If you want to explore sorting algorithms in more depth, perform an internet search for additional sorting algorithms.

Bubble Sort

A *bubble sort* involves a nested loop whereby each element of an array is compared with the elements to the right of the given element. If an array element is less than the current element, the values are interchanged ("swapped"), which means that the contents of the array will eventually be sorted from smallest to largest value.

Here is an example:

```
arr1 = np.array([40, 10, 30, 20]);
Item = 40;
Step 1: 40 > 10 so switch these elements:
arr1 = np.array([10, 40, 30, 20]);
Item = 40;
Step 2: 40 > 30 so switch these elements:
arr1 = np.array([10, 30, 40, 20]);
Item = 40;
Step 3: 40 > 20 so switch these elements:
arr1 = np.array([10, 30, 20, 40]);
```

As you can see, the smallest element is in the left-most position of the array `arr1`. Now repeat this process by comparing the second position (which is index 1) with the right-side elements.

```
arr1 = np.array([10, 30, 20, 40]);
Item = 30;
Step 4: 30 > 20 so switch these elements:
arr1 = np.array([10, 20, 30, 40]);
Item = 30;
Step 4: 30 < 40 so do nothing
```

As you can see, the smallest elements two elements occupy the first two positions in the array `arr1`. Now repeat this process by comparing the third position (which is index 2) with the right-side elements.

```
arr1 = np.array([10, 20, 30, 40]);
Item = 30;
Step 4: 30 < 40 so do nothing
```

The array `arr1` is now sorted in increasing order (in a left-to-right fashion). If you want to reverse the order so that the array is sorted in decreasing order (in a left-to-right fashion), simply replace the ">" operator with the "<" operator in the preceding steps.

Listing 4.4 displays the contents of the `BubbleSort.java` that illustrates how to perform a bubble sort on an array of numbers.

LISTING 4.4: BubbleSort.java

```
public class BubbleSort
{
   public static void main(String[] args)
   {
      int[] arr1 = new int[]{40, 10, 30, 20};
      System.out.print("Initial: ");
      for(int i=0; i<arr1.length; i++) {
        System.out.print(arr1[i]+" ");
      }
      System.out.println("");

      for(int i=0; i<arr1.length-1; i++) {
        for(int j=i+1; j<arr1.length; j++) {
          if(arr1[i] > arr1[j]) {
            int temp = arr1[i];
            arr1[i] = arr1[j];
            arr1[j] = temp;
          }
        }
      }

      System.out.print("Sorted:  ");
      for(int i=0; i<arr1.length; i++) {
```

```
          System.out.print(arr1[i]+" ");
        }
        System.out.println("");
    }
}
```

You can manually perform the code execution in Listing 4.4 to convince yourself that the code is correct. (Hint: it's the same sequence of steps that you saw earlier in this section). Launch the code in Listing 4.4 and you will see the following output:

```
initial: [40 10 30 20]
sorted:  [10 20 30 40]
```

Find Anagrams in a List of Words

Recall that the variable `word1` is an anagram of `word2` if `word2` is a permutation of `word1`. Listing 4.5 displays the contents of the `Anagrams2.java` that illustrates how to check if two words are anagrams of each other.

LISTING 4.5: Anagrams2.java

```
import java.util.Arrays;

public class Anagrams2
{
   public static String myStringSort(String str)
   {
      char[] arr = str.toCharArray();
      Arrays.sort(arr);
      String sorted = String.valueOf(arr);
      return sorted;
   }

   public static Boolean isAnagram(String str1, String str2)
   {
     String sorted1 = myStringSort(str1);
     String sorted2 = myStringSort(str2);
     return (sorted1.equals(sorted2));
   }

   public static void main(String[] args)
   {
      String[] words = new String[]{"abc","evil","Z","cab",
      "live","xyz","zyx","bac"};
     System.out.println("INITIAL: "+Arrays.toString(words));
      System.out.println();

      for(int i=0; i<words.length-1; i++) {
        for(int j=i+1; j<words.length; j++) {
```

```
            Boolean result = isAnagram(words[i], words[j]);
            if(result == true)
              System.out.println(words[i]+" and "+words[j]+"
              are anagrams");
            //else
            //  System.out.println(words[i]+" and
            "+words[j]+" are NOT anagrams");
          }
        }
      }
    }
```

Listing 4.5 starts with the definition of the static method myStringSort() that sorts the letters in an input string and then returns a string consisting of the sorted letters. The next portion of Listing 4.5 defines the static method `isAnagram()` that invokes the myStringSort() method in order to compare its two parameters. The sorted versions of the two parameters are compared, and the return statement returns either true or false depending on whether the sorted strings are equal or unequal, respectively.

The next portion of Listing 4.5 contains a main() method that initializes the variable `words` as an array of strings and then displays the contents of `words`. The next portion of code consists of a nested loop that compares each string in words with the subsequent strings in the variable words. For example, the first string is compared with the second through last (right-most) string, the second string is compared with the third through last (right-most) string, and so forth.

Each pair of strings in the nested loop is compared by invoking the method is_anagran() and a message is printed if any pair strings is indeed a pair of anagrams. Now launch the code in Listing 4.5 and you will see the following output:

```
=> Initial words:
['abc', 'evil', 'Z', 'cab', 'live', 'xyz', 'zyx', 'bac']

abc  and  cab  are anagrams
abc  and  bac  are anagrams
evil  and  live  are anagrams
cab  and  bac  are anagrams
xyz  and  zyx  are anagrams
```

SELECTION SORT

Listing 4.6 displays the contents of `SelectionSort.java` that illustrates how to perform a selection sort on an array of numbers.

LISTING 4.6: SelectionSort.java

```
public class SelectionSort
{
```

```
    public static void main(String[] args)
    {
       int[] arr1 = new int[]{64, 25, 12, 22, 11};

       System.out.print("Initial: ");
       for(int i=0; i<arr1.length; i++) {
         System.out.print(arr1[i]+" ");
       }
       System.out.println ("");

       //Traverse through all array elements
       for(int i=0; i<arr1.length; i++) {
         //Find the minimum element in remaining unsorted array
         int min_idx = i;
         for(int j=i+1; j<arr1.length; j++) {
           if(arr1[min_idx] > arr1[j])
             min_idx = j;
         }

         //Swap the found minimum element with the first element
         int temp = arr1[i];
         arr1[i] = arr1[min_idx];
         arr1[min_idx] = temp;
       }

       System.out.print("Sorted:  ");
       for(int i=0; i<arr1.length; i++) {
         System.out.print(arr1[i]+" ");
       }
       System.out.println();
    }
}
```

Listing 4.6 starts by initializing the array `arr1` with some integers, followed by a loop that iterates through the elements of `arr1`. During each iteration of this loop, another inner loop compares the current array element with each element that appears to the right of the current array element. If any of those elements is smaller than the current array element, then the index position of the former is maintained in the variable `min_idx`. After the inner loop has completed execution, the current element is "swapped" with the small element (if any) that has been found via the following code snippet:

```
arr1[i], arr1[min_idx] = arr1[min_idx], arr1[i]
```

In the preceding snippet, `arr1[i]` is the "current" element and `arr1[min_idx]` is element (to the right of index `i`) that is smaller than `arr1[i]`. If these two values are the same, then the code snippet swaps `arr1[i]` with itself. Now launch the code in Listing 4.6 and you will see the following output:

```
Initial: 64 25 12 22 11
Sorted:  11 12 22 25 64
```

INSERTION SORT

Listing 4.7 displays the contents of `InsertionSort.java` that illustrates how to perform a selection sort on an array of numbers.

LISTING 4.7: InsertionSort.java

```
public class InsertionSort
{
   public static void insertionSort(int[] arr1)
   {
     // Traverse through 1 to len(arr1)
     for(int i=1; i<arr1.length; i++) {
        int key = arr1[i];

        //Move elements of arr1[0..i-1], that are
        //greater than key, to one position ahead
        //of their current position
        int j = i-1;
        while((j >=0) && (key < arr1[j])) {
           arr1[j+1] = arr1[j];
           j -= 1;
        }
        arr1[j+1] = key;
        //System.out.println("New order: "+arr1);
     }
   }

   public static void main(String[] args)
   {
      int[] arr1 = new int[]{12, 11, 13, 5, 6};
      System.out.print("Initial:  ");
      for(int i=0; i<arr1.length; i++) {
         System.out.print(arr1[i]+" ");
      }
      System.out.println();

      insertionSort(arr1);
      System.out.print("Sorted:    ");
      for(int i=0; i<arr1.length; i++) {
         System.out.print(arr1[i]+" ");
      }
      System.out.println();
   }
}
```

Listing 4.7 starts with the function `insertionSort()` that contains a loop that iterates through the elements of the array `arr1`. During each iteration of this loop, the variable key is assigned the value of the element of array `arr1` whose index value is the loop variable `i`. Next, a `while` loop shifts a set of elements to the right of index `j`, as shown here:

```
j = i-1
while j >=0 and key < arr1[j] :
  arr1[j+1] = arr1[j]
  j -= 1
arr1[j+1] = key
```

For example, after the first iteration of the inner while loop we have the following output:

```
Initial:    [12, 11, 13, 5, 6]
New order: [11, 12, 13, 5, 6]
```

The second iteration of the inner loop does not produce any changes, but the third iteration shifts some of the array elements, at which point we have the following output:

```
Initial:    [12, 11, 13, 5, 6]
New order: [11, 12, 13, 5, 6]
New order: [11, 12, 13, 5, 6]
New order: [5, 11, 12, 13, 6]
```

The final iteration of the outer loop results in an array with sorted elements. Launch the code in Listing 4.7 and you will see the following output:

```
Initial:  12 11 13 5 6
Sorted:     5 6 11 12 13
New order: [5, 6, 11, 12, 13]
```

COMPARISON OF SORT ALGORITHMS

A bubble sort is rarely used: it's most effective when the data values are already almost sorted. A selection sort is used infrequently: while this algorithm is effective for very short lists, the insertion sort is often superior. An insertion sort is useful if the data are already almost sorted, or if the list is very short (e.g., at most 50 items).

Among the three preceding algorithms, only insertion sort is used in practice, typically as an auxiliary algorithm in conjunction with other more sophisticated sorting algorithms (e.g., quicksort or merge sort).

MERGE SORT

A *merge sort* involves merging two arrays of sorted values. In the following subsections you will see three different ways to perform a merge sort. The first code sample involves a third array, whereas the second and third code samples do not require a third array. Moreover, the third code sample involves one `while` loop whereas the second code sample involves a pair of nested loops, which means that the third code sample is simpler and also more memory efficient.

Merge Sort With a Third Array

The simplest way to merge two arrays involves copying elements from those two arrays to a third array, as shown here:

```
     A          B            C

  +-----+    +-----+     +-----+
  |  20 |    |  50 |     |  20 |   A
  |  80 |    |  70 |     |  50 |   B
  | 200 |  + | 100 |  =  |  70 |   B
  | 300 |    +-----+     |  80 |   A
  | 500 |                | 100 |   B
  +-----+                | 200 |   A
                         | 300 |   A
                         | 500 |   A
                         +-----+
```

The right-most column in the preceding diagram lists the array (either `A` or `B`) that contains each number. As you can see, the order `ABBABAAA` switches between array `A` and array `B`. However, the final three elements are from array `A` because all the elements of array `B` have been processed.

Two other possibilities exist: array `A` is processed and `B` still has some elements, or both `A` and `B` have the same size. Of course, even if `A` and `B` have the same size, it's still possible that the final sequence of elements are from a single array.

For example, array `B` is longer than array `A` in the example below, which means that the final values in array `C` are from `B`:

```
A = [20,80,200,300,500]
B = [50,70,100]
```

The following example involves array `A` and array `B` with the same length:

```
A = [20,80,200]
B = [50,70,300]
```

The next example also involves array `A` and array `B` with the same length, but all the elements of `A` are copied to `B` and then all the elements of `B` are copied to `C`:

```
A = [20,30,40]
B = [50,70,300]
```

Listing 4.8 displays the contents of `MergeSort1.java` that illustrates how to perform a merge sort on two arrays of numbers.

LISTING 4.8: MergeSort1.java

```
public class MergeSort1
{
   public static void MergeSort1(){}
```

```
public static int[] MergeItems(int[] items1, int[]
items2, int[] items3)
{
  int ndx1 = 0, ndx2 = 0, ndx3 = 0;

  // => always add the smaller element first:
  while(ndx1 < items1.length && ndx2 < items2.length)
  {
    //System.out.println(
    //"items1 data:"+items1[ndx1]+" items2
    data:"+items2[ndx2]);

    int data1 = items1[ndx1];
    int data2 = items2[ndx2];
    if(data1 < data2) {
      //System.out.println("adding data1: "+data1);
      items3[ndx3] = data1;
      ndx1 += 1;
    } else {
      //System.out.println("adding data2: "+data2);
      items3[ndx3] = data2;
      ndx2 += 1;
    }
    ndx3 += 1;
  }

  // append any remaining elements of items1:
  while(ndx1 < items1.length) {
      //System.out.println("MORE items1: "+items1[ndx1]);
      items3[ndx3] = items1[ndx1];
      ndx1 += 1;
  }

  // append any remaining elements of items2:
  while(ndx2 < items2.length) {
      //System.out.println("MORE items2: "+items2[ndx2]);
      items3[ndx3] = items2[ndx2];
      ndx2 += 1;
  }

  return items3;
}

public static void displayItems(int[] items)
{
   for(int item : items)
     System.out.print(item+" ");
   System.out.println("");
}
```

```
   public static void main(String[] args)
   {
      int[] items1 = new int[]{20, 30, 50, 300};
      int[] items2 = new int[]{80, 100, 200};
      int[] items3 = new int[items1.length+items2.length];

      // display the initial and merged lists:
      System.out.println("First sorted array:");
      displayItems(items1);
      System.out.println("");

      System.out.println("Second sorted array:");
      displayItems(items2);
      System.out.println("");

      System.out.println("Merged array:");
      items3 = mergeItems(items1, items2, items3);
      displayItems(items3);
   }
}
```

Listing 4.8 defines the static methods `mergeItems()` and `displayItems()` that perform the merge operation on two sorted arrays and display the initial arrays and merged array, respectively.

The `mergeItems()` method contains three loops. The first loop iterates through the elements of `items1` and `items2` and performs a comparison: whichever element is smaller is added to the array `items3`. The second loop adds any remaining elements in items1 to `arr3`, and the third loop adds any remaining elements in `items2` to `arr3`. The `displayItems()` method method displays the contents of the array that is passed to this method.

The next portion of Listing 4.8 defines the method main() that initializes the sorted arrays `items1` and `items2` with integer and allocates storage for the "target" array items3. The next three (short) code blocks display the contents of `items1` and `items2`, and then invoke the method `mergeItems()` to merge these two arrays, after which the merged contents are displayed. Now launch the code in Listing 4.8 and you will see the following output:

```
First sorted array:
20 30 50 300

Second sorted array:
80 100 200

Merged array:
20 30 50 80 100 200 300
```

Merge Sort Without a Third Array

Listing 4.9 displays the contents of `MergeSort2.java` that illustrates how to perform a merge sort on two *sorted* arrays without using a third array.

LISTING 4.9: MergeSort2.java

```
public class MergeSort2
{
   public static void displayItems(int[] items)
   {
      for(int item : items)
        System.out.print(item+" ");
      System.out.println("");
   }

   public static void mergeArrays()
   {
     int[] items1 = new int[] {20, 30, 50, 300, 0, 0, 0, 0};
      int[] items2 = new int[] {80, 100, 200,999};

      System.out.println("=> Merge items2 into items1");
      System.out.println("Sorted array items1:");
      displayItems(items1);
      System.out.println("");
      System.out.println("Sorted array items2:");
      displayItems(items2);
      System.out.println("");

      int ndx1 = 0, ndx2 = 0, ndx3 = 0;
      int last1 = 4; // do not count the 0 values in items1

     // merge elements of items2 into items1:
     while(ndx2 < items2.length)
     {
        //System.out.println(
        //"items1 data: "+items1[ndx1]+" items2 data:
        "+items2[ndx2]);
        int data1 = items1[ndx1];
        int data2 = items2[ndx2];

        // skip over elements in items1 that
        // are < their counterpart in items2:
        while(data1 < data2 )
        {
           if (ndx1 < items1.length-1) {
              //System.out.println("incrementing ndx1:
              "+ndx1);
              ndx1  += 1;
              data1 = items1[ndx1];
```

```
                } else {
                   //System.out.println("past length of array
                   //items1");
                   break;
                }
             }

             // right-shift ONE element in items1
             // to insert an element from items2:
             for(int idx3=last1; idx3 >ndx1; idx3--) {
               //System.out.println("shift "+items1[idx3]+" to
               //the right");
               items1[idx3] = items1[idx3-1];
             }

             // insert data2 into items1:
             items1[ndx1] = data2;
             ndx1   = 0;
             ndx2  += 1;
             last1 += 1;
             // System.out.println("=> shifted items1:
             //"+items1);
          }

          System.out.println("Merged sorted array items3:");
          displayItems(items1);
       }

       public static void main(String[] args)
       {
          mergeArrays();
       }
    }
```

Listing 4.9 defines the static methods `mergeArrays()` and `displayItems()` that perform the merge operation on two sorted arrays and display the initial arrays and merged array, respectively.

The `mergeArrays()` initializes the sorted arrays `items1` and `items2` and then displays their contents by invoking the `displayItems()` method with each array. The next portion of mergeArrays() contains an outer loop that iterates through the values of `items2`. During each iteration, an inner loop is executed that increments the index `ndx1` for `items1` until the value in the `items1` array is greater than the current value in the items2 array. At this point another loop performs a right-shift of the elements in the `items1` array in order to insert the current element of the `items2` array. The outer loop terminates when all of the elements in items2 have been processed: remember that the purpose of the code is to insert the elements from the array `items2` into the array `items1` so that the result is a sorted array.

The next portion of Listing 4.9 defines the method `main()` that invokes the `mergeArrays()` method. Launch the code in Listing 4.9 and you will see the following output:

```
=> Merge items2 into items1
Sorted array items1:
20 30 50 300 0 0 0 0

Sorted array items2:
80 100 200 999

System.out.println("Merged sorted array items3:");
20 30 50 80 100 200 300 999
```

Merge Sort: Shift Elements From End of Lists

In this scenario we assume that matrix `A` has enough uninitialized elements at the end of the matrix in order to accommodate all the values of matrix `B`, as shown in Figure 4.1.

```
      A              B              A

  +-----+        +-----+        +-----+
  |  20 |        |  50 |        |  20 |   A
  |  80 |        |  70 |        |  50 |   B
  | 200 |   +    | 100 |   =    |  70 |   B
  | 300 |        +-----+        |  80 |   A
  | 500 |                       | 100 |   B
  | xxx |                       | 200 |   A
  | xxx |                       | 300 |   A
  | xxx |                       | 500 |   A
  +-----+                       +-----+
```

FIGURE 4.1. Merging two arrays.

HOW DOES QUICKSORT WORK?

The *quicksort* algorithm uses a divide-and-conquer approach to sort an array of elements. The key idea involves selecting an item in a given list as the *pivot* item (which can be any item in the list) that is used for partitioning the given list into two sublists and then recursively sorting the two sublists.

Due to the recursive nature of this algorithm, each recursive invocation results in smaller sublists. Hence, the sublists eventually reach the base cases where the sublists have either 0 or 1 elements (which are obviously sorted).

Another key point: one sublist contains values that are less than the pivot item, and the other sublist contains values that are greater than the pivot item.

In the ideal case, both sublists are approximately the same length. This results in a binary-like splitting of the sublists, which involves `log N` invocations of the quicksort algorithm, where N is the number of elements in the list.

There are several points to keep in mind regarding the quicksort. First, the case in which the two sublists are approximately the same length is the more efficient case. However, this case involves a prior knowledge of the data distribution in the given list in order to achieve optimality.

Second, if the list contains values that are close to randomly distributed, in which case the first value or the last value are common choices for the pivot item. Third, quicksort has its worst performance when the values in a list are already sorted. In this scenario, select the pivot item in one of the following ways:

- Select the middle item in the list.
- Select the median of the first, middle, and last items in the list.

Quicksort Code Sample

Listing 4.10 displays the contents of `QuickSort.java` that illustrates how to perform a quicksort on an array of numbers.

LISTING 4.10: QuickSort.java

```
public class QuickSort
{
   public static void sort(int[] values) {
     quicksort(values);
   }

   public static void quicksort(int[] arr) {
     if(arr.length == 0)
       return;
     else
       quicksort(arr, 0, arr.length - 1);
   }

   // Sort interval [lo, hi] inplace recursively
   public static void quicksort (int[] arr, int lo, int hi)
   {
     if (lo < hi) {
       int splitPoint = partition(arr, lo, hi);
       quicksort(arr, lo, splitPoint);
       quicksort(arr, splitPoint + 1, hi);
     }
   }

   // Performs Hoare partition algorithm for quicksort
   public static int partition(int[] arr, int lo, int hi)
   {
     int pivot = arr[lo];
```

```
      int i = lo - 1;
      int j = hi + 1;

      while (true) {
        do {
          i += 1;
        }
        while (arr[i] < pivot);

        do {
          j -= 1;
        }
        while (arr[j] > pivot);

        if (i < j) swap(arr, i, j);
        else return j;
      }
    }

    // Swap two elements
    public static void swap(int[] arr, int i, int j)
    {
      int tmp = arr[i];
      arr[i] = arr[j];
      arr[j] = tmp;
    }

    public static void main(String[] args)
    {
       int[] array = new int[]{10, 4, 6, 4, 8, -13, 2, 3};

       System.out.println("Initial array:");
       for (int num : array )
         System.out.print(num+" ");
       System.out.println();

       sort(array);
       System.out.println("Sorted  array:");
       for (int num : array )
         System.out.print(num+" ");
       System.out.println();
    }
}
```

Listing 4.10 contains the definition of the following static methods, followed by the `main()` method:

- `sort()`
- `quicksort()`
- `partition()`
- `swap()`

The method `sort()` invokes the overloaded method `quicksort()`, which in turn invokes the second `quicksort()` method if the parameter `arr` (which is an array) has non-zero length.

Next, the second `quicksort()` method contains conditional logic that is executed when the parameter `lo` is less than the parameter `hi`, which involves three code snippets. The first code snippet initializes the value `splitPoint` with the result of executing the method `partition()`, which is discussed later. The second code snippet recursively invokes the current `quicksort()` method with one set of arguments, followed by another recursive invocation of executes the current `quicksort()` method with a second set of arguments.

The next portion of Listing 4.10 is the method `partition()` that performs the "heavy lifting" of the quicksort algorithm. This method starts by initializing the variables pivot, `i`, and `j`, with the values of `arr[lo]`, `lo-1`, and `hi+1`, respectively, which are the parameters of the current method.

The remaining code in the method `partition()` consists of a `while` loop that contains two `do-while` loops and an `if/else` statement. The first `while` loop increments the value of `i` as long as the value of `arr[i]` is less than the value of `pivot`, whereas the second `while` loop *decrements* the value of `j` as long as the value of `arr[i]` is greater than the value of `pivot`. Next, the `if/else` statement invokes the method `swap()` if `i` is less than `j`; otherwise, this method returns the value of `j`. The method `swap()` is a standard method for swapping the values of two variables.

The next portion of Listing 4.10 is the `main()` method that initializes the variable `array` with a list of eight integer values, followed by a code block that displays the contents of the variable `array`.

The next portion of the main() method invokes the `sort()` method, followed by another code block that displays the contents of the sorted array. Launch the code in Listing 4.10 and you will see the following output:

```
Sorted array: [1, 5, 7, 8, 9, 10]
```

SHELL SORT

The *Shell sort* (by Donald L. Shell) is similar to bubble sort: both algorithms compare pairs of elements and then swap them if they are out of order. However, unlike bubble sort, Shell sort does not compare adjacent elements until the last iteration. Instead, it first compares elements that are widely separated, shrinking the size of the gap with each pass. In this way, it deals with elements that are significantly out of place early on, reducing the amount of work that later passes must do.

Listing 4.11 displays the contents of `ShellSort.java` that illustrates how to implement the Shell sort algorithm in `Java`.

LISTING 4.11: ShellSort.java

```
public class ShellSort
{
   public static int[] shellSort(int num, int[] arr)
   {
     int gap = (int)(num/2);

     while(gap > 0) {
       //for i in range(gap,num-1):
       for(int i=gap; i<num; i++) {
         int j = i-gap;

         while(j >= 0 && arr[j] > arr[j+gap]) {
           // swap arr[j] and arr[j+gap]
           int temp = arr[j];
           arr[j] = arr[j+gap];
           arr[j+gap] = temp;
           j = j-gap ;
         }
       }
       gap = (int)(gap/2);
     }

     return arr;
   }

   public static void main(String[] args)
   {
      int num = 6;
      int[] arr = new int[]{50,20,80,-100,500,200};

      System.out.println("Original:");
      for (int num1 : arr )
        System.out.print(num1+" ");
      System.out.println();

      int[] result = shellSort(num, arr);
      System.out.println("Sorted:");
      for (int num2 : result )
        System.out.print(num2+" ");
      System.out.println();
   }
}
```

Listing 4.11 defines the static method `shellSort()` that contains the Shell sort algorithm. The first line of code initializes the integer-valued variable `gap` as the midpoint of the array `arr` that contains integer values. Next, a `while` loop executes as long as the variable `gap` is greater than 0. During each iteration, a nested `for` loop and another `while` loop is executed. The `for` loop initializes the loop variable `i` with the value of `gap` and increments the value of `i` up to the value `num-1`, and also initializes the variable `j` with the value `i-gap`, as shown here:

```
for(int i=gap; i<num; i++) {
  int j = i-gap;
```

Next, the nested `while` loop swaps the elements in index positions `j` and j+gap and replaces the value of `j` with `j-gap`. When the nested `for` loop has completed execution, the value of `gap` divided by 2, and the outer loop executes again.

The next portion of Listing 4.11 defines the `main()` method that initializes the array `arr` with six integer values, and initializes the variable `num` with the length of `arr`. The remaining portion of the `main()` method displays the values of area, invokes the `shellSort()` method, and then displays the sorted values in the array `arr`. Launch the code in Listing 4.11 and you will see the following output:

```
Original:
50 20 80 -100 500 200
Sorted:
-100 20 50 80 200 500
```

TASK: SORTED ARRAYS AND THE SUM OF TWO NUMBERS

Listing 4.12 displays the contents of the `PairSumTarget.java` that illustrates how to determine whether or not a sorted array contains the sum of two specified numbers.

LISTING 4.12: PairSumTarget.java

```
// given two numbers num1 and num2 in a sorted array,
// determine whether or not num1+num2 is in the array
public class PairSumTarget
{
   public static void checkSum(int[] arr1, int num1, int num2)
   {
     int ndx1 = 0;
     int ndx2 = arr1.length-1;
     //System.out.println("Anum1: "+num1+" num2: "+num2);

     while(arr1[ndx1] < num1)
       ndx1 += 1;
     //System.out.println("Bndx1: "+ndx1+" ndx2: "+ndx2);

     while(arr1[ndx2] > num2)
       ndx2 -= 1;
     //System.out.println("Cndx1: "+ndx1+" ndx2: "+ndx2);

     System.out.println("Contents of array:");
```

```
      for (int num3 : arr1 )
        System.out.print(num3+" ");
      System.out.println("\n");

      int sum = num1+num2;
      System.out.println("Checking for the sum => "+sum);

      //System.out.println("num1: "+num1+" num2: "+num2);
      //System.out.println("ndx1: "+ndx1+" ndx2: "+ndx2);
      // NOTE: arr1[ndx1] >= num1 AND arr1[ndx2] >= sum

      if(arr1[ndx1]+arr1[ndx2] == sum) {
        System.out.println("=> FOUND the sum "+sum);
        System.out.println("indexes: "+ndx1+" and "+ndx2);
      } else {
        System.out.println("=> SUM NOT FOUND: "+sum);
      }
      System.out.println();
    }

    public static void main(String[] args)
    {
       int[] arr1 = new int[]{20,50,100,120,150,200,250,300};
       int num1 = 60;
       int num2 = 90;
       checkSum(arr1,num1,num2);

       arr1 = new int[] {20,50,100,120,150,200,250,300};
       num1 = 60;
       num2 = 100;
       checkSum(arr1,num1,num2);

       arr1 = new int[]{3,3};
       num1 = 3;
       num2 = 3;
       checkSum(arr1,num1,num2);
    }
}
```

Listing 4.12 defines the static method `checkSum()` that calculates the sum of pairs of array elements to determine whether or not that sum is equal to a given value. This calculation is performed in a loop that *increments* a lower index `ndx1` while the array value in `arr` is less than `num1`, and then *decrements* an upper index `ndx2` while the array value is larger than `num2`. If the sum of `arr[ndx1]` and `arr[ndx2]` equals the value in the variable `sum`, then a target sum has been found, and an appropriate message is displayed.

The next portion of Listing 4.12 defines the main() method that initializes the array `arr1` with a sorted list of integers and then displays the contents of `arr1`. The remaining portion of the `main()` method initializes the variables `num1` and `num2` and then invokes the `checkSum()` method. Now launch the code in Listing 4.12 and you will see the following output:

```
Original:
50 20 80 -100 500 200
Sorted:
-100 20 50 80 200 500
```

SUMMARY

This chapter started with search algorithms such as linear search and binary search (iterative and recursive). You also saw a comparison between a linear search and a binary search.

Next, you larked about some well-known sorting algorithms, such as the bubble sort, selection sort, insertion sort, and the merge sort. Finally, you learned about the quicksort that you can perform on an array of items.

CHAPTER 5

LINKED LISTS

This chapter introduces you to a data structure called a *linked list*, which includes singly linked lists, doubly linked lists, and circular lists. Linked lists support various operations, such as create, traverse, insert, and delete operations. This chapter shows you how to implement linked lists and how to perform some operations on linked lists. Chapter 6 contains various task-related code samples that involve more than just the basic operations on linked lists. Keep in mind that algorithms for linked lists often involve recursion, which is discussed in Chapter 2.

The first part of this chapter introduces you to linked lists, followed by examples of performing various operations on singly linked lists, such as creating and displaying the contents of linked lists, as well as updating nodes and deleting nodes in a singly linked list.

The second part of this chapter introduces you to doubly linked lists, followed by examples of performing various operations on doubly linked lists, which are the counterpart to the code samples for singly linked lists.

One other point to keep in mind for this chapter as well as other chapters: the code samples provide a solution that prefers clarity over optimization. In addition, many code samples contain "commented out" `print()` statements. If the code samples confuse you, uncomment the `print()` statements in order to trace the execution path of the code: doing so can make the code easier to follow and also save you a lot of time. Indeed, after you have read each code sample and you fully understand the code, try to optimize the code, or perhaps use a different algorithm, which will enhance your problem-solving skills as well as increase your confidence level.

One other detail regarding the naming convention for variables in this chapter as well as Chapter 6. For code samples that involve linked lists, the "root" node is labeled ROOT, the "current" node is labeled CURR, and the "last" node is labeled LAST. This style is intended to draw your attention to those variables whenever the associated list elements are created or modified. However, there are other coding styles that you can adopt in which the preceding variables are written in lowercase instead of uppercase.

Finally, if you are already comfortable with linked lists, then you might already be prepared for Chapter 6 that contains well-known tasks that pertain to linked lists.

TYPES OF DATA STRUCTURES

This section introduces you to the concept of linear data structures (stacks and queues) and nonlinear data structures (trees and graphs), some of which are discussed in Chapters 5, 6, and 7. Those chapters indicate the tasks for which those data structures are well suited (e.g., inserting new elements or finding existing elements). Moreover, you will see `Java` code samples for each of those data structures that perform operations such as how to insert, update, and remove elements.

Linear Data Structures

Linear data structures are data structures whose elements occur in sequential memory locations or are logically connected, such as stacks and queues. As you will see in Chapter 6, stacks are last-in-first-out (LIFO) data structures, which means that the last element inserted is the first element removed. Moreover, operations are performed from one end of the stack. A real-life counterpart to a stack is an elevator that has a single entrance that is also the lone exit.

In Chapter 7, you will learn about queues, which are FIFO data structures, which means that the first element inserted is the first element removed. By contrast with a stack, insert operations are performed at the so-called "front" of the queue and delete operations are performed at the "rear" of the queue. A real-life counterpart to a queue is a line of people waiting to purchase tickets to a movie.

Nonlinear Data Structures

Nonlinear data structures are data structures whose elements are not sequential, such as trees and graphs. As you will see in Chapter 7, trees have a single node called the root node that has no parent node, whereas every other node has exactly one parent node. Two nodes are related if there is an edge connecting the two nodes. In fact, the nodes in a tree have a hierarchical relationship. As you will see in Chapter 7, trees can have undirected edges, directed edges, and cycles.

A graph is a generalization of a tree, and nodes in a graph can have multiple parent nodes. Moreover, a graph does not have a root node: instead, a graph can have a "source" and a "sink" that are somewhat analogous to a "start" node and an "end" node. This designation appears in graphs that represent transport networks in which edges can have weights assigned to them. Think of trucks that transport food or other commodities from a warehouse (the "source") and have multiple routes to reach their destination (the "sink").

DATA STRUCTURES AND OPERATIONS

In this book, the data structure for singly linked lists is a custom `Node` structure that consists of the following:

- a key/value pair
- a pointer to next node

The data structure for doubly linked lists is a custom `Node` that consists of the following:

- a key/value pair
- a pointer to next node
- a pointer to previous node

The data structure for a stack and also for a queue is a `Java` list.

The data structure for trees is a custom node that consists of the following:

- a key/value pair
- a pointer to left node
- a pointer to right node

The data structure for a hash table is a `Java` dictionary. The preceding structures are based on simplicity; however, please keep in mind that other structures are possible as well. For example, you can use an array to implement a list, a stack, a queue, and even a tree.

Operations on Data Structures

As you will see later in this chapter as well as in other chapters, the operations on these data structures usually involve the following:

- insert (which includes append)
- delete
- search
- update (an existing element)
- check for empty structure
- check for full structure (queues and stacks)

WHAT ARE SINGLY LINKED LISTS?

Although arrays are useful data structures, one of their disadvantages is that the size or the number of elements in an array must be known in advance. One alternative that does not have this disadvantage is a "linked list," which is the topic of this chapter.

A singly linked list is a collection of data elements, which are commonly called nodes. Nodes contain two things:

- a value that is stored in the node
- the location of the next node (called its successor)

By way of analogy, think of a conga line of dances: each dancer is a node, and the dancer places his or her hands on the hips of another dancer: the latter dancer is the location of the next node.

Unlike arrays, linked lists are dynamically created on an as-needed basis. Moreover, the preceding analogy makes the following point clear: there is a "last" node that does not have a next node. Thus, the last node in a list has `None` as its successor (i.e., the next node).

In general, a node in a singly linked list can be one of the following three types:

- the "root" node
- an intermediate node
- the last node (no next element)

Of course, when a linked list is empty, then the nodes in the preceding list are not initialized.

Tradeoffs for Linked Lists

Every data structure has tradeoffs, which is to say, advantages and disadvantages. In particular, linked lists have the following advantages:

- easy to append an element to a list
- easy to insert an element to a list
- easy to delete an element to a list
- the number of items in the list is not required in advance

Unlike arrays, the operations in the preceding bullet items do not require "shifting over" the elements of an array: only the pointers are updated.

However, linked lists do have disadvantages:

- more difficult (time-consuming) to search for an element
- a larger amount of memory is required

Thus, arrays work better when the number of elements is known in advance and there are no insertions or deletions (only updates), whereas linked lists work better when the number of data elements is not known in advance.

Unlike arrays, the elements of a linked list can be stored in noncontiguous memory locations: the value of the next field provides the location of the next node in the linked list (except for the `LAST` node that has `None` as its next node).

SINGLY LINKED LISTS: CREATE AND APPEND OPERATIONS

Linked lists support several operations, including insert (add a new node), delete (an existing node), update (change the value of an existing node), and traverse (list all the nodes). The following subsections discuss the preceding operations in more detail.

A Node Class for Singly Linked Lists

Listing 5.1 displays the contents of `SLNode.java` that illustrates how to define a simple `Java` class that defines the Node class for a single node in a linked list.

LISTING 5.1: SLNode.py

```
public class SLNode
{
   static class Node {
     String data = "";
     public Node(String data) {
       this.data = data;
       Node next = null;
     }
   }

   public static void main(String[] args)
   {
      Node node1 = new Node("Jane");
      Node node2 = new Node("Dave");
      Node node3 = new Node("Stan");
      Node node4 = new Node("Alex");

      System.out.println("node1.data: "+node1.data);
      System.out.println("node2.data: "+node2.data);
      System.out.println("node3.data: "+node3.data);
      System.out.println("node4.data: "+node4.data);
   }
}
```

Listing 5.1 is straightforward: it initializes the variables msg and num with the specified values. Launch the following command from the command line and you will see the following output:

```
node1.data: Jane
node2.data: Dave
node3.data: Stan
node4.data: Alex
```

Java Code for Appending a Node

Listing 5.2 displays the contents of `AppendSLNode.java` that illustrates a better way to create a linked list and append nodes to that list.

LISTING 5.2: AppendSLNode.java

```
public class AppendSLNode
{
   static class Node {
     String data = "";
     Node next = null;

     public Node(String data) {
       this.data = data;
       this.next = null;
     }
   }

   public static Node[] appendNode(Node ROOT, Node LAST,
   String item)
   {
     if(ROOT == null) {
       ROOT = new Node(item);
       //System.out.println("1Node:", ROOT.data);
     } else {
       if(ROOT.next == null) {
         Node NEWNODE = new Node(item);
         LAST = NEWNODE;
         ROOT.next = LAST;
         //System.out.println("2Node:", NEWNODE.data);
       } else {
         Node NEWNODE = new Node(item);
         LAST.next = NEWNODE;
         LAST = NEWNODE;
         //System.out.println("3Node: "+NEWNODE.data);
       }
     }

     Node[] results = new Node[2];
     results[0] = ROOT;
     results[1] = LAST;
     return results;
   }

   public static void main(String[] args)
   {
      Node[] results = new Node[2];
      Node ROOT  = null;
      Node LAST  = null;

      // append items to list:
      String[] items = new String[]{"Stan", "Steve",
      "Sally", "Alex"};
      for(String item : items) {
        results = appendNode(ROOT, LAST, item);
```

```
            ROOT = results[0];
            LAST = results[1];
        }

        // display items in list:
        Node CURR = ROOT;
        while(CURR != null) {
            System.out.println("Node: "+CURR.data);
            CURR = CURR.next;
        }
    }
}
```

Listing 5.2 defines a `Node` class as before, followed by the `Java` method `appendNode()` that contains the logic for initializing a singly linked list and also for appending nodes to that list.

Launch the code in Listing 5.2 from the command line and you will see the following output:

```
Node: Stan
Node: Steve
Node: Sally
Node: Alex
```

FINDING A NODE IN A LINKED LIST

Listing 5.3 displays the contents of `FindSLNode2.java` that illustrates how to find a node in a linked list.

LISTING 5.3: FindSLNode2.java

```
public class FindSLNode2
{
   static class Node {
     String data = "";
     Node next = null;

     public Node(String data) {
       this.data = data;
       this.next = null;
     }
   }

   public static Node[] appendNode(Node ROOT, Node LAST,
   String item)
   {
     if(ROOT == null) {
       ROOT = new Node(item);
       //System.out.println("1Node:", ROOT.data);
     } else {
```

```
      if(ROOT.next == null) {
        Node NEWNODE = new Node(item);
        LAST = NEWNODE;
        ROOT.next = LAST;
        //System.out.println("2Node:", NEWNODE.data);
      } else {
        Node NEWNODE = new Node(item);
        LAST.next = NEWNODE;
        LAST = NEWNODE;
        //System.out.println("3Node:", NEWNODE.data);
      }
    }

    Node[] results = new Node[2];
    results[0] = ROOT;
    results[1] = LAST;
    return results;
  }

  public static void findItem(Node ROOT,String item)
  {
    Boolean found = false;
    Node CURR = ROOT;
    System.out.println("=> Search for: "+item);
    while (CURR != null) {
      System.out.println("Checking: "+CURR.data);
      if(CURR.data == item) {
        System.out.println("=> Found "+item);
        found = true;
        break;
      } else {
        CURR = CURR.next;
      }
    }

    if(found == false) {
      System.out.println("* "+item+" not found *");
    }
    System.out.println("-------------------\n");
  }

  public static void displayList(Node ROOT)
  {
     Node CURR = ROOT;
     while (CURR != null) {
        System.out.print(CURR.data+" ");
        CURR = CURR.next;
     }
     System.out.println("\n");
  }
```

```
    public static void main(String[] args)
    {
       Node[] results = new Node[2];
       Node ROOT   = null;
       Node LAST   = null;

       String[] items = new String[]{"Stan", "Steve",
       "Sally", "Alex"};
       for(String item : items) {
         results = appendNode(ROOT, LAST, item);
         ROOT = results[0];
         LAST = results[1];
       }

       System.out.println("List of Items:");
       displayList(ROOT);

       for(String item : items)
         find_item(ROOT,item);
    }
}
```

Listing 5.3 defines the static methods appendNode(), findItem(), and displayList() that append nodes to a linked list, find items in a linked list, and display the contents of a linked list, respectively.

The appendNode() method contains the same code that you saw in a previous example, and the findItem() iterates through the linked list and compares the contents of each node with a given string. An appropriate message is displayed depending on whether or not the given string was found in the linked list.

The main() method starts by initializing the string-based array items with a list of strings, followed by a loop that iterates through the elements of items and invokes the appendNode() method in order to append each string to the linked list. The final portion of the main() method displays the contents of the linked list, followed by a loop that iterates through the elements in the items variable and checks whether or not each element is present in the linked list. Launch the code in Listing 5.3 from the command line and you will see the following output:

```
List of Items:
Stan Steve Sally Alex

=> Search for: Stan
Checking: Stan
=> Found Stan
-------------------

=> Search for: Steve
Checking: Stan
```

```
Checking: Steve
=> Found Steve
-------------------

=> Search for: Sally
Checking: Stan
Checking: Steve
Checking: Sally
=> Found Sally
-------------------

=> Search for: Alex
Checking: Stan
Checking: Steve
Checking: Sally
Checking: Alex
=> Found Alex
-------------------
```

Listing 5.3 works fine for a small number of nodes. For linked lists that contain more nodes, we need a scalable way to construct a list of nodes, which is discussed in the next section.

APPENDING A NODE IN A LINKED LIST

When you create a linked list, you must *always* check if the root node is empty: if so, then you create a root node, otherwise you append the new node to the last node. Let's translate the preceding sentence into pseudocode that describes how to add a new element to a linked list:

```
Let ROOT be the root node (initially NULL) of the linked
list
Let LAST be the last node (initially NULL) of the linked
list
Let NEW be a new node and let NEW->next = NULL

// decide where to insert the new node:
if (ROOT == NULL)
{
   ROOT = NEW;
   LAST = NEW;
}
else
{
   LAST->next = NEW;
   LAST = NEW;
}
```

The last node in a linked list always points to a NULL element.

Finding a Node in a Linked List (Method 2)

Listing 5.4 displays the contents of `FindSLNode.java` that illustrates how to find a node in a linked list.

LISTING 5.4: FindSLNode.java

```
public class FindSLNode
{
   static class Node {
     String data = "";
     Node next = null;
     public Node(String data) {
       this.data = data;
       this.next = null;
     }
   }

   public static void findItem(Node ROOT, String item)
   {
      Boolean found = false;
      Node CURR = ROOT;
      System.out.println("=> Search for: "+item);

      while (CURR != null) {
        System.out.println("Checking: "+CURR.data);
        if(CURR.data == item) {
          System.out.println("=> Found "+item);
          found = true;
          break;
        } else {
          CURR = CURR.next;
        }
      }

      if(found == false)
        System.out.println("* "+item+" not found *");
      System.out.println("-------------------\n");
   }

   public static Node[] appendNode(Node ROOT, Node LAST,
   String item)
   {
     if(ROOT == null) {
       ROOT = new Node(item);
       //print("1Node:", ROOT.data);
     } else {
       if(ROOT.next == null) {
         Node NEWNODE = new Node(item);
         LAST = NEWNODE;
         ROOT.next = LAST;
```

```
          //print("2Node:", NEWNODE.data);
        } else {
          Node NEWNODE = new Node(item);
          LAST.next = NEWNODE;
          LAST = NEWNODE;
          //print("3Node:", NEWNODE.data)
        }
      }

      Node[] results = new Node[2];
      results[0] = ROOT;
      results[1] = LAST;
      return results;
    }

    public static void main(String[] args)
    {
       Node[] results = new Node[2];

       Node ROOT  = null;
       Node LAST  = null;

       Node node1 = new Node("Jane");
       ROOT  = node1;

       Node node2 = new Node("Dave");
       LAST = node2;
       ROOT.next = LAST;

       Node node3 = new Node("Stan");
       LAST.next = node3;
       LAST  = node3;

       Node node4 = new Node("Alex");
       LAST.next = node4;
       LAST  = node4;

       ROOT  = null;
       LAST  = null;

       String[] items = new String[]{"Stan", "Steve",
       "Sally", "Alex"};
       for(String item : items) {
         results = appendNode(ROOT, LAST, item);
         ROOT = results[0];
         LAST = results[1];
       }

       for(String item : items)
         findItem(ROOT,item);
    }
}
```

Listing 5.4 is straightforward: it initializes the variables `msg` and `num` with the specified values. Launch the code in Listing 5.4 from the command line and you will see the following output:

```
=> Search for: Stan
Checking: Jane
Checking: Dave
Checking: Stan
=> Found Stan
-------------------

=> Search for: Steve
Checking: Jane
Checking: Dave
Checking: Stan
Checking: Alex
* Steve not found *
-------------------

=> Search for: Sally
Checking: Jane
Checking: Dave
Checking: Stan
Checking: Alex
* Sally not found *
-------------------

=> Search for: Alex
Checking: Jane
Checking: Dave
Checking: Stan
Checking: Alex
=> Found Alex
```

SINGLY LINKED LISTS: UPDATE AND DELETE OPERATIONS

In the previous section, you saw how to create a linked list, display its contents, and search for a specific node. In this section, you will learn how to update a node in a linked list as also how to delete a node in a linked list.

UPDATING A NODE IN A SINGLY LINKED LIST

The following pseudocode explains how to search for an element, and update its contents if the element is present in a linked list:

```
CURR = ROOT
Found = False
OLDDATA = "something old";
NEWDATA = "something new";
```

```
if (ROOT == NULL)
{
   print("* EMPTY LIST *");
}

while (CURR != NULL)
{
   if(CURR->data == OLDDATA)
   {
      print("found node with value",OLDDATA);
      CURR->data = NEWDATA;
   }

   if(Found == True) { break; }

   PREV = CURR;
   CURR = CURR->next;
}
```

Java Code to Update a Node

Listing 5.5 displays the contents of `UpdateSLNode.java` that illustrates how to update a node in a linked list.

LISTING 5.5: UpdateSLNode.java

```
public class UpdateSLNode
{
   static class Node {
     String data = "";
     Node next = null;

     public Node(String data) {
       this.data = data;
       this.next = null;
     }
   }

   public static Node[] appendNode(Node ROOT, Node LAST,
   String item)
   {
     if(ROOT == null) {
       ROOT = new Node(item);
       //print("1Node:", ROOT.data);
     } else {
       if(ROOT.next == null) {
         Node NEWNODE = new Node(item);
         LAST = NEWNODE;
         ROOT.next = LAST;
         //print("2Node:"+ NEWNODE.data);
```

```
    } else {
      Node NEWNODE = new Node(item);
      LAST.next = NEWNODE;
      LAST = NEWNODE;
      //print("3Node:", NEWNODE.data)
    }
  }

  Node[] results = new Node[2];
  results[0] = ROOT;
  results[1] = LAST;
  return results;
}

public static void main(String[] args)
{
   Node[] results = new Node[2];
   Node ROOT  = null;
   Node LAST  = null;

   String[] items = new String[]{"Stan", "Steve",
   "Sally", "Alex"};
   for(String item : items) {
     results = appendNode(ROOT, LAST, item);
     ROOT = results[0];
     LAST = results[1];
   }

   // display items in list:
   System.out.println("=> list items:");
   Node CURR = ROOT;
   while(CURR != null) {
     System.out.println("Node: "+CURR.data);
     CURR = CURR.next;
   }
   System.out.println();

   // update item in list:
   String curr_val = "Alex";
   String new_val  = "Alexander";
   Boolean found = false;

   CURR = ROOT;
   while(CURR != null)
   {
     if(CURR.data == curr_val) {
       System.out.println("Found:  "+CURR.data);
       CURR.data = new_val;
       System.out.println("Updated: "+CURR.data);
       found = true;
       break;
```

```
            } else {
              CURR = CURR.next;
            }
          }

          if(found == false)
            System.out.println("* Item "+curr_val+" not in list *");
      }
   }
```

Listing 5.5 defines a `Node` class as before, followed by the `Java` method `appendNode()` that contains the logic for initializing a singly linked list and also for appending nodes to that list.

Now launch the code in Listing 5.5 from the command line and you will see the following output:

```
=> list items:
Node: Stan
Node: Steve
Node: Sally
Node: Alex

Found:   Alex
Updated: Alexander
```

DELETING A NODE IN A LINKED LIST

The following pseudocode explains how to search for an element, and delete the element if it is present in a linked list:

```
CURR = ROOT
PREV = ROOT
item = <node-value>
Found = False

if (ROOT == NULL)
{
   print("* EMPTY LIST *");
}

while (CURR != NULL)
{
   if(CURR.data == item)
   {
      print("found node with value",item);

      Found = True
      if(CURR == ROOT)
      {
         ROOT = CURR.next // the list is now empty
      }
```

```
        else
        {
           PREV.next = CURR.next;
        }
     }

     if(found == True) { break; }

     PREV = CURR;
     CURR = CURR.next;
  }

  return ROOT
```

JAVA CODE FOR DELETING A NODE

Listing 5.6 displays the contents of `DeleteSLNode.java` that illustrates how to delete a node in a linked list. This code sample is longer than the code samples in the previous sections because the code needs to distinguish between deleting the root node versus a nonroot node.

LISTING 5.6: DeleteSLNode.java

```
public class DeleteSLNode
{
   static class Node {
     String data = "";
     Node next = null;
     public Node(String data) {
       this.data = data;
       this.next = null;
     }
   }

   public static Node[] appendNode(Node ROOT, Node LAST,
   String item)
   {
     if(ROOT == null) {
       ROOT = new Node(item);
     //System.out.println("1Node: "+ ROOT.data);
     } else {
       if(ROOT.next == null) {
         Node NEWNODE = new Node(item);
         LAST = NEWNODE;
         ROOT.next = LAST;
       //System.out.println("2Node: "+ NEWNODE.data);
       } else {
         Node NEWNODE = new Node(item);
         LAST.next = NEWNODE;
         LAST = NEWNODE;
```

```
      //System.out.println("3Node: "+ NEWNODE.data);
      }
    }

  //System.out.println("NEW ROOT: "+ROOT.data);
  //System.out.println("NEW LAST: "+LAST.data);
    Node[] results = new Node[2];
    results[0] = ROOT;
    results[1] = LAST;
    return results;
  }

  public static Node deleteItem(Node ROOT, String item)
  {
    Node PREV = ROOT;
    Node CURR = ROOT;
    Boolean found = false;

    System.out.println("=> searching for item: "+item);
    while (CURR != null) {
       if(CURR.data == item) {
          //System.out.println("=> Found node with value:
          "+item);
          found = true;

          if(CURR == ROOT) {
            ROOT = CURR.next;
          //System.out.println("NEW ROOT");
          } else {
          //System.out.println("REMOVED NON-ROOT");
            PREV.next = CURR.next;
          }
       }

       if(found == true)
         break;

       PREV = CURR;
       CURR = CURR.next;
    }

    if(found == false)
      System.out.println("* Item "+item+" not in linked
      list *\n");
    else
      System.out.println("* Removed "+item+" from linked
      list *\n");

    return ROOT;
  }
```

```
  public static void displayItems(Node ROOT)
  {
    System.out.println("=> Items in Linked List:");
    Node CURR = ROOT;
    while(CURR != null) {
      System.out.println("Node: "+CURR.data);
      CURR = CURR.next;
    }
    System.out.println();
  }

  public static Node initializeList(Node ROOT)
  {
     Node LAST = ROOT;
     // initialize linked list:
     String[] items = new String[]{"Stan", "Steve",
     "Sally", "Alex"};

     Node[] results = new Node[2];
     for(String item : items) {
       results = appendNode(ROOT, LAST, item);
       ROOT = results[0];
       LAST = results[1];
     }

     return ROOT;
  }

  public static void main(String[] args)
  {
     Node ROOT = null;
     Node LAST = null;
     Node[] results = new Node[2];

     // construct linked list:
     ROOT = initializeList(ROOT);

     // display items in linked list:
     displayItems(ROOT);

     // remove item from linked list:
     ROOT = deleteItem(ROOT, "Alex");

     displayItems(ROOT);
  }
}
```

Listing 5.6 defines a `Node` class as before, followed by the `Java` method `appendNode()` that contains the logic for initializing a singly linked list and also for appending nodes to that list.

Launch the code in Listing 5.6 from the command line and you will see the following output:

```
=> Items in Linked List:
Node: Stan
Node: Steve
Node: Sally
Node: Alex

=> searching for item: Alex
* Removed Alex from linked list *

=> Items in Linked List:
Node: Stan
Node: Steve
Node: Sally
```

JAVA CODE FOR A CIRCULAR LINKED LIST

The only structural difference between a singly linked list and a circular linked list is that the "last" node in the latter has a "next" node equal to the initial (root) node. Operations on circular linked lists are the same as operations on singly linked lists. However, the algorithms for singly linked lists need to be modified in order to accommodate circular linked lists.

Listing 5.7 displays the contents of `CircularSLNode.java` that illustrates how to delete a node in a linked list. This code sample is longer than the code samples in the previous sections because the code needs to distinguish between deleting the root node versus a nonroot node.

LISTING 5.7: CircularSLNode.java

```
public class CircularSLNode
{
   static class Node {
     String data = "";
     Node next = null;

     public Node(String data) {
       this.data = data;
       this.next = null;
     }
   }

/*
The code follows these cases for inserting a node:
   NULL    (empty list)
   x       (one node)
   x<->x   (two nodes)
   x y z x (multiple nodes)
*/
```

```
public static Node appendNode(Node ROOT, String item)
{
  System.out.println("=> processing item: "+item);

  if(ROOT == null) {
    ROOT = new Node(item);
    ROOT.next = ROOT;
    //System.out.println("1NEW initialized Root: "+ROOT.
    data);
  } else {
    if(ROOT.next == ROOT) {
      //System.out.println("Found SINGLE NODE Root:
      "+ROOT.data);
      Node NEWNODE = new Node(item);
      ROOT.next = NEWNODE;
      NEWNODE.next = ROOT;
      //System.out.println("Root: "+ROOT.data+" NEWNODE:
      "+NEWNODE.data);
    } else {
      // traverse the list to find the node prior to ROOT:
      Node LAST = ROOT;
      LAST = LAST.next;
      while(LAST.next != ROOT) {
         LAST = LAST.next;
      }
      Node NEWNODE = new Node(item);
      NEWNODE.next = ROOT;
      LAST.next = NEWNODE;
     //System.out.println("3Node NEW LAST: "+NEWNODE.data);
    }
  }
  //System.out.println("Returning ROOT: "+ROOT.data);
  return ROOT;
}

public static void displayList(Node ROOT)
{
   Node CURR = ROOT;
   if( ROOT == null ) {
      System.out.print("* empty list *");
      return;
   }

   System.out.print(CURR.data+" ");
   CURR = CURR.next;

   // display items in list:
   while(CURR != ROOT) {
     System.out.print(CURR.data+" ");
     CURR = CURR.next;
```

```
        }
        System.out.println();
    }

    public static void main(String[] args)
    {
        Node ROOT = null;

        String[] items = new String[]{"Stan","Steve","Sally",
        "Alex","Nancy","Sara"};

        // append items to list:
        for(String item : items) {
          ROOT = appendNode(ROOT, item)
        }

        System.out.println("Contents of circular list:");
        displayList(ROOT);
    }
}
```

Listing 5.7 defines a `Node` class as before, followed by the `Java` function `appendNode()` that contains the logic for initializing a singly linked list and also for appending nodes to that list.

Launch the code in Listing 5.7 from the command line and you will see the following output:

```
1ROOT: Stan

=> processing item: Stan
=> processing item: Steve
=> processing item: Sally
=> processing item: Alex
=> processing item: Nancy
=> processing item: Sara
Contents of circular list:
Stan Steve Sally Alex Nancy Sara
```

JAVA CODE FOR UPDATING A CIRCULAR LINKED LIST

The only structural difference between a singly linked list and a circular linked list is that the "last" node in the latter has a "next" node equal to the initial (root) node. Operations on circular linked lists are the same as operations on singly linked lists. However, the algorithms for singly linked lists need to be modified in order to accommodate circular linked lists.

Listing 5.8 displays the contents of `CircularUpdateSLNode.java` that illustrates how to delete a node in a linked list. This code sample is longer than the code samples in the previous sections because the code needs to distinguish between deleting the root node versus a nonroot node.

LISTING 5.8: CircularUpdateSLNode.java

```
public class CircularUpdateSLNode
{
   static class Node {
     String data = "";
     Node next = null;

     public Node(String data) {
       this.data = data;
       this.next = null;
     }
   }

/*
The code follows these cases for inserting a node:
   NULL    (empty list)
   x       (one node)
   x<->x   (two nodes)
   x y z x (multiple nodes)
*/
   public static Node appendNode(Node ROOT, String item)
   {
     //System.out.println("=> processing item: "+item);

     if(ROOT == null) {
       ROOT = new Node(item);
       ROOT.next = ROOT;
       //System.out.println("1NEW initialized Root: "+ROOT.
       data);
     } else {
       if(ROOT.next == ROOT) {
         //System.out.println("Found SINGLE NODE Root:
         "+ROOT.data);
         Node NEWNODE = new Node(item);
         ROOT.next = NEWNODE;
         NEWNODE.next = ROOT;
         //System.out.println(
         //.     "Root: "+ROOT.data+" NEW: "+NEWNODE.data);
       } else {
         // traverse the list to find the node prior to ROOT:
         Node LAST = ROOT;
         LAST = LAST.next;
         while(LAST.next != ROOT) {
            LAST = LAST.next;
         }

         Node NEWNODE = new Node(item);
         NEWNODE.next = ROOT;
         LAST.next = NEWNODE;
```

```
            //System.out.println("3Node NEW LAST: "+NEWNODE.
            data);
          }
        }
        //System.out.println("Returning ROOT: "+ROOT.data);
        return ROOT;
    }

    public static Node updateItem(Node ROOT, String curr_
    val, String new_val)
    {
       if(ROOT == null) {
          System.out.println("=> empty list *");
          return null;
       }

       System.out.println("OLD: "+curr_val+" NEW: "+new_val);

       if(ROOT.next == ROOT) {
          System.out.println("ROOT VALUE: "+ROOT.data);
          if(curr_val.equals(ROOT.data)) {
             System.out.println(
               "Found data: "+curr_val+" new value: "+new_val);
             ROOT.data = new_val;
             return ROOT;
          }
       }

       Node CURR = ROOT;
       if(curr_val.equals(CURR.data)) {
          System.out.println(
             "Found data: "+curr_val+" new value: "+new_val);
          CURR.data = new_val;
          return ROOT;
       }

       // compare the contents of each node with the new string:
       Boolean Found = false;
       CURR = CURR.next;
       while(CURR != ROOT) {
         if(CURR.data == curr_val) {
           System.out.println(
             "Found data: "+curr_val+" new value: "+new_val);
           CURR.data = new_val;
           Found = true;
           break;
         } else {
           CURR = CURR.next;
         }
       }
```

```
    if(Found == false)
      System.out.println("* "+curr_val+" not found *");

    return ROOT;
  }

  public static void displayList(Node ROOT)
  {
    Node CURR = ROOT;
    if( ROOT == null ) {
       System.out.println("* empty list *");
       return;
    }

    System.out.print("LIST: ");
    System.out.print(CURR.data+" ");
    CURR = CURR.next;

    // display items in list:
    while(CURR != ROOT) {
      System.out.print(CURR.data+" ");
      CURR = CURR.next;
    }
    System.out.println();
  }

  public static void main(String[] args)
  {
    Node ROOT = null;

    String[] items =
      new String[]{"Stan","Steve","Sally","Alex","Nancy",
      "Sara"};

    // append items to list:
    for(String item : items) {
      ROOT = appendNode(ROOT, item);
    }

    System.out.println("Contents of circular list:");
    displayList(ROOT);
    System.out.println();

    // list of update strings:
    String[] replace1 =
        new String[]{"Stan","Sally","Sara","Tomas"};
    String[] replace2 =
        new String[]{"Joe","Sandra","Rebecca","Theo"};

    for(int i=0; i<replace1.length; i++) {
```

```
            // update value of a list item:
            String curr_val = replace1[i];
            String new_val  = replace2[i];
            ROOT = update_item(ROOT, curr_val, new_val);
            display_list(ROOT);
            System.out.println();
        }
    }
}
```

Listing 5.8 defines a `Node` class as before, followed by the `Java` method `appendNode()` that contains the logic for initializing a singly linked list and also for appending nodes to that list.

Launch the code in Listing 5.8 from the command line and you will see the following output:

```
Contents of circular list:

Contents of circular list:
LIST: Stan Steve Sally Alex Nancy Sara

OLD VALUE: Stan NEW VALUE: Joe
Found data: Stan new value: Joe
LIST: Joe Steve Sally Alex Nancy Sara

OLD VALUE: Sally NEW VALUE: Sandra
Found data: Sally new value: Sandra
LIST: Joe Steve Sandra Alex Nancy Sara

OLD VALUE: Sara NEW VALUE: Rebecca
Found data: Sara new value: Rebecca
LIST: Joe Steve Sandra Alex Nancy Rebecca

OLD VALUE: Tomas NEW VALUE: Theo
* Tomas not found *
LIST: Joe Steve Sandra Alex Nancy Rebecca
```

WORKING WITH DOUBLY LINKED LISTS (DLL)

A doubly linked list is a collection of data elements, which are commonly called nodes. Given a node, it contains three items:

1. `a value that is stored in the node`
2. `the location of the` **`next`** `node (called its successor)`
3. `the location of the` **`previous`** `node (called its predecessor)`

Using the same "conga line" analogy as described in the previous section about singly linked lists, each person knows is touching the "next" node with one hand and is touching the "previous" node with the other hand (not the best analogy, but you get the idea).

Operations on doubly linked lists are the same as the operations on singly linked lists; however, there are two pointers (the successor and the predecessor) to update instead of just one (the successor).

A Node Class for Doubly Linked Lists

Listing 5.9 displays the contents of the `Java` file `DLNode.java` that illustrates how to define a simple `Java` class that represents a single node in a linked list.

LISTING 5.11: DLNode.java

```
public class DLNode
{
   public String data = "";
   public String prev = "";
   public String next = "";

   public DLNode( String data )
   {
     this.data = data;
     this.prev = "";
     this.next = "";
   }

   public static void main( String[] args)
   {
      DLNode node1 = new DLNode("Jane");
      DLNode node2 = new DLNode("Dave");
      DLNode node3 = new DLNode("Stan");
      DLNode node4 = new DLNode("Alex");

      System.out.println("node1.data: "+node1.data);
      System.out.println("node2.data: "+node2.data);
      System.out.println("node3.data: "+node3.data);
      System.out.println("node4.data: "+node4.data);
   }
}
```

Listing 5.9 is straightforward: it defines the `Java` class `DLNode` and then creates four such nodes. The final portion of Listing 5.9 displays the contents of the four nodes. Launch the following command from the command line and you will see the following output:

```
node1.data: Jane
node2.data: Dave
node3.data: Stan
node4.data: Alex
```

Once again, a node in a doubly linked list can be one of the following three types:

```
1)the "root" node
2)an intermediate node
3)the last node (no next element)
```

Of course, when a linked list is empty, then the nodes in the preceding list are all `None`.

APPENDING A NODE IN A DOUBLY LINKED LIST

When you create a linked list, you must always check if the root node is empty: if so, then you create a root node, otherwise you append the new node to the last node. Let's translate the preceding sentence into pseudocode that describes how to add a new element to a linked list.

```
Let ROOT be the root node (initially NULL) of the linked
list
Let LAST be the last node (initially NULL) of the linked
list
Let NEW be a new node with NEW->next = NULL and NEW->prev =
NULL

# decide where to insert the new node:
if (ROOT == NULL)
{
   ROOT = NEW;
   ROOT->next = NULL;
   ROOT->prev = NULL;

   LAST = NEW;
   LAST->next = NULL;
   LAST->prev = NULL;
}
else
{
   NEW->prev = LAST;
   LAST->next = NEW;
   NEW->next = NULL;
   LAST = NEW;
}
```

The last node in a doubly linked list always points to a `null` element.

Java Code for Appending a Node

Listing 5.10 displays the contents of `AppendDLNode.java` that illustrates a better way to create a linked list and append nodes to that list.

LISTING 5.10: AppendDLNode.java

```
public class AppendDLNode
{
   static class DLNode {
     String data = "";
     DLNode next = null;
     DLNode prev = null;

     public DLNode(String data) {
       this.data = data;
       this.next = null;
       this.prev = null;
     }
   }

   public static DLNode constructList()
   {
      DLNode ROOT = null;
      DLNode LAST = null;
      DLNode[] results = new DLNode[2];

      String[] items = new String[]{"Jane", "Dave", "Stan",
      "Alex"};

      for(String item : items) {
         results = appendNode(ROOT, LAST, item);
         ROOT = results[0];
         LAST = results[1];
      }

      return ROOT;
   }

   public static DLNode[] appendNode(DLNode ROOT, DLNode
   LAST, String item)
   {
     if(ROOT == null) {
       ROOT = new DLNode(item);
       ROOT.next = ROOT;
       ROOT.prev = ROOT;
       LAST = ROOT;
       //System.out.println("1DLNode:", ROOT.data);
     } else {
       if(ROOT.next == null) {
         DLNode NEWNODE = new DLNode(item);
         LAST.next = NEWNODE;
         NEWNODE.prev = LAST;
         LAST = NEWNODE;
         //System.out.println("2DLNode:", NEWNODE.data);
       } else {
```

```
            DLNode NEWNODE = new DLNode(item);
            LAST.next = NEWNODE;
            NEWNODE.prev = LAST;
            NEWNODE.next = null;
            LAST = NEWNODE;
            //System.out.println("3DLNode:", NEWNODE.data);
          }
        }

        DLNode[] results = new DLNode[2];
        results[0] = ROOT;
        results[1] = LAST;
        return results;
      }

      public static void displayList(DLNode ROOT)
      {
         System.out.println("Doubly Linked List:");
         DLNode CURR = ROOT;
         while (CURR != null) {
           System.out.println(CURR.data);
           CURR = CURR.next;
         }
      }

      public static void main( String[] args)
      {
         DLNode ROOT = null;
         ROOT = constructList();
         displayList(ROOT);
      }
}
```

Listing 5.10 defines a `Node` class as before, followed by the method `appendNode()` that contains the logic for initializing a singly linked list and also for appending nodes to that list.

Launch the code in Listing 5.10 from the command line and you will see the following output:

```
1ROOT: Stan
2ROOT: Stan
2LAST: Steve
3Node: Sally
3Node: Alex

=> list items:
Node: Stan
Node: Steve
Node: Sally
Node: Alex
```

```
=> update list items:
Found data in LAST: Alex

=> list items:
Node: Stan
Node: Steve
Node: Sally
Node: Alexander
```

Java Code for Inserting a New Root Node

Listing 5.11 displays the contents of the `Java` file `NewRootSLNode.java` that illustrates how to iterate through an array of strings and set each string as the new root node, which effectively creates a linked list in reverse order.

LISTING 5.11: NewRootSLNode.java

```
public class NewRootSLNode
{
   static class SLNode {
     String data = "";
     SLNode next = null;

     public SLNode(String data) {
       this.data = data;
       this.next = null;
     }
   }

   public static SLNode constructList(SLNode ROOT, SLNode LAST)
   {
      String[] items = new String[]{"Stan", "Steve",
      "Sally", "Alex"};

      System.out.println("Initial items:");
      for(String item : items) {
         System.out.print(item+" ");
      }
      System.out.println("\n");

      for(String item : items) {
        ROOT = new_root(ROOT, item);
      }
      return ROOT;
   }

   public static SLNode newRoot(SLNode ROOT, String item)
   {
     if(ROOT == null) {
       ROOT = new SLNode(item);
```

```
            ROOT.next = null;
            //print("1SLNode:", ROOT.data);
        } else {
            SLNode NEWNODE = new SLNode(item);
            NEWNODE.next = ROOT;
            ROOT = NEWNODE;
        }
        return ROOT;
    }

    public static void displayList(SLNode ROOT)
    {
        System.out.println("Reversed Linked List:");
        SLNode CURR = ROOT;
        while(CURR != null) {
          System.out.print(CURR.data+" ");
          CURR = CURR.next;
        }
        System.out.println("\n");
    }

    public static void main(String[] args)
    {
        SLNode ROOT = null;
        SLNode LAST = null;

        // construct SL list:
        ROOT = constructList(ROOT, LAST);

        displayList(ROOT);
    }
}
```

Listing 5.11 defines a `Node` class as before, followed by the definition of the methods constructList(), newRoot(), and displayList() that construct a linked list, insert a new root node, and display the contents of the constructed linked list, respectively.

The `constructList()` displays the elements of the string array `items`, followed by a loop that iterates through the elements of the items array. During each iteration, the `newRoot()` method is invoked with a string that is used to create a new node, which in turn is set as the new root node of the linked list. The `displayList()` method contains the same code that you have seen in previous examples.

The `main()` method is very simple: the ROOT and LAST nodes are initialized as null nodes and then the constructList() method is invoked. The only other code snippet invokes the `displayList()` method that displayed the elements in the (reversed) list. Launch the code in Listing 5.11 from the command line and you will see the following output:

```
Initial items:
Stan Steve Sally Alex

Reversed Linked List:
Alex Sally Steve Stan
```

Java Code for Inserting an Intermediate Node

Listing 5.12 displays the contents of `NewInterSLNode.java` that illustrates a better way to create a linked list and append nodes to that list.

LISTING 5.12: NewInterSLNode.java

```
public class NewInterSLNode
{
   static class Node {
     String data = "";
     Node next = null;

     public Node(String data) {
       this.data = data;
       this.next = null;
     }
   }

/*
The code follows these cases for deleting a node:
   NULL  (empty list)
   x     (one node)
   x x   (two nodes)
   x x x (multiple nodes)
*/
   public static void insertItem(Node ROOT, int position,
   String data)
   {
      if(ROOT == null) {
         System.out.println("* empty list *");
      }

      int count=0;
      Node CURR = ROOT;
      Node PREV = CURR;

      while (CURR != null) {
        if(count == position) {
           System.out.println("=> "+data+" after position
           "+position);
           Node NEWNODE = new Node(data);
           if(CURR.next == null) {
              CURR.next = NEWNODE;
```

```
        } else {
           NEWNODE.next = CURR.next;
           CURR.next = NEWNODE;
        }
        break;
     }
     count += 1;
     PREV = CURR;
     CURR = CURR.next;
   }

   if(count < position) {
      System.out.println(position+" is beyond end of list");
  }
}

public static Node[] appendNode(Node ROOT, Node LAST,
String item)
{
  if(ROOT == null) {
    ROOT = new Node(item);
    //print("1Node:", ROOT.data);
  } else {
    if(ROOT.next == null) {
      Node NEWNODE = new Node(item);
      LAST = NEWNODE;
      ROOT.next = LAST;
      //print("2Node:", NEWNODE.data);
    } else {
      Node NEWNODE = new Node(item);
      LAST.next = NEWNODE;
      LAST = NEWNODE;
      //print("3Node:", NEWNODE.data)
    }
  }

  Node[] results = new Node[2];
  results[0] = ROOT;
  results[1] = LAST;
  return results;
}

public static void displayList(Node ROOT)
{
   Node CURR = ROOT;
   while (CURR != null) {
      System.out.print(CURR.data+" ");
      CURR = CURR.next;
   }
   System.out.println("\n");
}
```

```
    public static void main(String[] args)
    {
       Node[] results = new Node[2];
       Node ROOT  = null;
       Node LAST  = null;

       String[] items =
                new String[]{"Stan", "Steve", "Sally", "Alex"};
       for(String item : items) {
         results = appendNode(ROOT, LAST, item);
         ROOT = results[0];
         LAST = results[1];
       }

       System.out.println("Initial list:");
       displayList(ROOT);

       String[] insert_items =
                new String[]{"MIKE", "PASTA", "PIZZA", "CODE"};
       int[] insert_indexes  = new int[]{0, 3, 8, 2};

       int pos = 0;
       for(String new_item : insert_items) {
          System.out.println("Inserting: "+new_item);
          insertItem(ROOT, insert_indexes[pos++], new_item);
          System.out.println("Updated list:");
          displayList(ROOT);
       }
    }
}
```

Listing 5.12 defines a `Node` class as before, followed by the `Java` method `appendNode()` that contains the logic for initializing a singly linked list and also for appending nodes to that list.

Now launch the code in Listing 5.12 from the command line and you will see the following output:

```
Initial list:
Stan Steve Sally Alex

Inserting: MIKE
=> MIKE after position 0
Updated list:
Stan MIKE Steve Sally Alex

Inserting: PASTA
=> PASTA after position 3
Updated list:
Stan MIKE Steve Sally PASTA Alex
```

```
Inserting: PIZZA
8 is beyond end of list
Updated list:
Stan MIKE Steve Sally PASTA Alex

Inserting: CODE
=> CODE after position 2
Updated list:
Stan MIKE Steve CODE Sally PASTA Alex
```

Traversing the Nodes in a Doubly Linked List

The following pseudocode explains how to traverse the elements of a linked list:

```
CURR = ROOT

while (CURR != NULL)
{
   print("contents:",CURR->data);
   CURR = CURR->next;
}

If (ROOT == NULL)
{
   print("* EMPTY LIST *");
}
```

Updating a Node in a Doubly Linked List

The following pseudocode explains how to search for an element, and update its contents if the element is present in a linked list:

```
CURR = ROOT
Found = False
OLDDATA = "something old";
NEWDATA = "something new";

If (ROOT == NULL)
{
   print("* EMPTY LIST *");
}

while (CURR != NULL)
{
   if(CURR->data = OLDDATA)
   {
      print("found node with value",OLDDATA);
      CURR->data = NEWDATA;
   }

   if(Found == True) { break; }
```

```
        PREV = CURR;
        CURR = CURR->next;
    }
```

As you now know, some operations on doubly linked lists involve updating a single pointer and other operations involve updating two pointers.

Java Code to Update a Node

Listing 5.13 displays the contents of `UpdateDLNode.java` that illustrates how to update a node in a linked list.

LISTING 5.13: UpdateDLNode.java

```
public class UpdateDLNode
{
   static class DLNode {
     String data = "";
     DLNode next = null;
     DLNode prev = null;

     public DLNode(String data) {
       this.data = data;
       this.next = null;
       this.prev = null;
     }
   }

   public static DLNode constructList()
   {
      DLNode ROOT = null;
      DLNode LAST = null;
      DLNode[] results = new DLNode[2];

      String[] items = new String[]{"Jane", "Dave", "Stan",
      "Alex"};

      for(String item : items) {
         results = appendNode(ROOT, LAST, item);
         ROOT = results[0];
         LAST = results[1];
      }

      return ROOT;
   }

   public static DLNode[] appendNode(DLNode ROOT, DLNode
   LAST, String item)
   {
     if(ROOT == null) {
```

```
        ROOT = new DLNode(item);
        ROOT.next = ROOT;
        ROOT.prev = ROOT;
        LAST = ROOT;
        //System.out.println("1DLNode:", ROOT.data);
      } else {
        if(ROOT.next == null) {
          DLNode NEWNODE = new DLNode(item);
          LAST.next = NEWNODE;
          NEWNODE.prev = LAST;
          LAST = NEWNODE;
          //System.out.println("2DLNode:", NEWNODE.data);
        } else {
          DLNode NEWNODE = new DLNode(item);
          LAST.next = NEWNODE;
          NEWNODE.prev = LAST;
          NEWNODE.next = null;
          LAST = NEWNODE;
          //System.out.println("3DLNode:", NEWNODE.data);
        }
      }

      DLNode[] results = new DLNode[2];
      results[0] = ROOT;
      results[1] = LAST;
      return results;
    }

    public static void displayList(DLNode ROOT)
    {
       System.out.println("Doubly Linked List:");
       DLNode CURR = ROOT;
       while (CURR != null) {
         System.out.println(CURR.data);
         CURR = CURR.next;
       }
       System.out.println();
    }

    public static void updateNode(DLNode ROOT, String item,
    String data)
    {
       Boolean found = false;
       System.out.println("Searching for "+item+":");

       DLNode CURR = ROOT;
       while (CURR != null) {
         if(CURR.data == item) {
           System.out.println("Replacing "+item+" with "+data);
            CURR.data = data;
            found = true;
```

```
            break;
         }
         CURR = CURR.next;
      }

      if(found == false) {
         System.out.println(item+" not found in linked list");
      }
      System.out.println();
   }

   public static void main( String[] args)
   {
      DLNode ROOT = null;
      ROOT = constructList();
      displayList(ROOT);

      updateNode(ROOT, "steve", "sally");
      updateNode(ROOT, "Dave", "Davidson");
   }
}
```

Listing 5.13 defines a `Node` class as before, followed by the `Java` method `appendNode()` that contains the logic for initializing a singly linked list and also for appending nodes to that list.

Now launch the code in Listing 5.13 from the command line and you will see the following output:

```
Doubly Linked List:
Jane
Dave
Stan
Alex

Searching for steve:
steve not found in linked list

Searching for Dave:
Replacing Dave with Davidson
```

DELETING A NODE IN A DOUBLY LINKED LIST

The following pseudocode explains how to search for an element, and delete the element if it is present in a linked list:

```
CURR = ROOT
PREV = ROOT
ANODE = <a-node-to-delete>
Found = False

If (ROOT == NULL)
```

```
{
   print("* EMPTY LIST *");
}

while (CURR != NULL)
{
   if(CURR->data = ANODE->data)
   {
      print("found node with value",ANODE->data);

      Found = True
      if(CURR == ROOT)
      {
         ROOT = NULL; // the list is now empty
      }
      else
      {
         PREV->next = CURR->next;
      }
   }

   if(Found == True) { break; }

   PREV = CURR;
   CURR = CURR->next;
}
```

Java Code to Delete a Node

Listing 5.14 displays the contents of `DeleteDLNode.java` that illustrates how to update a node in a doubly linked list.

LISTING 5.14: DeleteDLNode.java

```
public class DeleteDLNode
{
   static class DLNode {
     String data = "";
     DLNode next = null;
     DLNode prev = null;

     public DLNode(String data) {
       this.data = data;
       this.next = null;
       this.prev = null;
     }
   }

   public static DLNode constructList()
   {
```

```
    DLNode ROOT = null;
    DLNode LAST = null;
    DLNode[] results = new DLNode[2];

    String[] items =
         new String[]{"Jane","Dave","Stan","Alex","George
         ","Sara"};

    for(String item : items) {
       results = appendNode(ROOT, LAST, item);
       ROOT = results[0];
       LAST = results[1];
    }

    return ROOT;
 }

 public static DLNode[] appendNode(DLNode ROOT, DLNode
 LAST, String item)
 {
   if(ROOT == null) {
     ROOT = new DLNode(item);
     ROOT.next = ROOT;
     ROOT.prev = ROOT;
     LAST = ROOT;
     //System.out.println("1DLNode:", ROOT.data);
   } else {
     if(ROOT.next == null) {
       DLNode NEWNODE = new DLNode(item);
       LAST.next = NEWNODE;
       NEWNODE.prev = LAST;
       LAST = NEWNODE;
       //System.out.println("2DLNode:", NEWNODE.data);
     } else {
       DLNode NEWNODE = new DLNode(item);
       LAST.next = NEWNODE;
       NEWNODE.prev = LAST;
       NEWNODE.next = null;
       LAST = NEWNODE;
       //System.out.println("3DLNode:", NEWNODE.data);
     }
   }

   DLNode[] results = new DLNode[2];
   results[0] = ROOT;
   results[1] = LAST;
   return results;
 }

 public static void displayList(DLNode ROOT)
 {
```

```
        //System.out.println("Doubly Linked List:");
        DLNode CURR = ROOT;
        while (CURR != null) {
          System.out.print(CURR.data+" ");
          CURR = CURR.next;
        }
        System.out.println("\n");
     }

  /*
  The code follows these cases for deleting a node:
     NULL  (empty list)
     x     (one node)
     x x   (two nodes)
     x x x (multiple nodes)
  */
     public static DLNode deleteNode(DLNode ROOT, String item)
     {
        Boolean found = false;
        System.out.println("Searching for "+item+":");

        if(ROOT == null) {
          System.out.println("* empty list *");
          return ROOT;
        }

        DLNode CURR = ROOT;
        DLNode PREV = ROOT;

        while (CURR != null) {
          if(CURR.data == item) {
             // three cases for the matched node:
             // root node, middle node, or last node:
             found = true;
             System.out.println("Deleting node with "+item);

             if(CURR == ROOT) {
               // is the list a single node?
               if(ROOT.next == null) {
                 ROOT = null;
               } else {
                 ROOT = ROOT.next;
                 ROOT.prev = null;
               }
               //System.out.println("new root: "+ROOT.data);
             } else {
               DLNode NEXT = CURR.next;
               if(NEXT == null) {
                 //System.out.println("final node: "+CURR.data);
                  PREV.next = null;
               } else {
```

```
                //System.out.println("middle node: "+CURR.
data);
                PREV.next = NEXT;
                NEXT.prev = PREV;
             }
          }
          break;
        }
        PREV = CURR;
        CURR = CURR.next;
      }

      if(found == false) {
        System.out.println(item+" not found in linked list");
      }
      return ROOT;
   }

   public static void main( String[] args)
   {
      DLNode ROOT = null;
      ROOT = constructList();

      System.out.println("Initial list:");
      displayList(ROOT);

      String[] remove_names = new String[]{"Jane", "Stan",
      "Isaac", "Sara"};
      for(String name : remove_names) {
         ROOT = deleteNode(ROOT, name);
         System.out.println("Updated list:");
         displayList(ROOT);
      }
   }
}
```

Listing 5.14 defines a `Node` class as before, followed by the `Java` method `appendNode()` that contains the logic for initializing a doubly linked list and also for appending nodes to that list.

Launch the code in Listing 5.14 from the command line and you will see the following output:

```
Initial list:
Jane Dave Stan Alex George Sara

Searching for Jane:
Deleting node with Jane
Updated list:
Dave Stan Alex George Sara

Searching for Stan:
```

```
Deleting node with Stan
Updated list:
Dave Alex George Sara

Searching for Isaac:
Isaac not found in linked list
Updated list:
Dave Alex George Sara

Searching for Sara:
Deleting node with Sara
Updated list:
Dave Alex George
```

SUMMARY

This chapter started with a description of linked lists, along with their advantages and disadvantages. Then you learned how to perform several operations on singly linked lists, such as append, insert, delete, and update.

In addition, you learned about doubly linked lists, and how to perform the same operations on doubly linked lists that you performed on singly linked lists.

CHAPTER 6

LINKED LISTS

Chapter 5 introduced you to singly linked lists and doubly linked lists, and how to perform basic operations on those data structures. This chapter shows you how to perform a variety of tasks that involve more than the basic operations in the previous chapter.

The first part of this chapter contains code samples that add the numbers in a linked list in several different ways, followed by code samples that display the first k and the last k elements in a linked list.

The code samples in the second portion of this chapter reverse a singly linked list and remove duplicates. You will also see how to concatenate two lists and how to merge two ordered lists. In addition, you will learn how to find the middle element of a linked list. The final portion of this chapter shows you how to reverse a list and how to check if a linked list contains a palindrome.

Recall from the introduction to Chapter 5 that in code samples that involve linked lists, the "root" node is labeled ROOT, the "current" node is labeled CURR, and the "last" node is labeled LAST. This style is intended to draw your attention to those variables whenever the associated list elements are created or modified. However, there are other coding styles that you can adopt in which the preceding variables are written in lowercase instead of uppercase.

TASK: ADDING NUMBERS IN A LINKED LIST (1)

Listing 6.1 displays the contents of `SumSLNodes.java` that illustrates how to add numbers in a linked list.

LISTING 6.1: SumSLNodes.java

```
public class SumSLNodes
{
   static class Node {
     int data;
     Node next = null;
```

```
    public Node(int data) {
      this.data = data;
      this.next = null;
    }
  }

  public static Node[] appendNode(Node ROOT, Node LAST,
  int item)
  {
    Node[] results = new Node[2];

    if(ROOT == null) {
      ROOT = new Node(item);
      //System.out.println("1Node: "+ROOT.data);
    } else {
      if(ROOT.next == null) {
        Node NEWNODE = new Node(item);
        LAST = NEWNODE;
        ROOT.next = LAST;
        //System.out.println("2Node: "+NEWNODE.data);
      } else {
        Node NEWNODE = new Node(item);
        LAST.next = NEWNODE;
        LAST = NEWNODE;
        //System.out.println("3Node: "+NEWNODE.data);
      }
    }

    //return ROOT, LAST
    results[0] = ROOT;
    results[1] = LAST;
    return results;
  }

  public static void main(String[] args)
  {
     Node ROOT  = null;
     Node LAST  = null;
     Node[] results = new Node[2];
     int[] items = new int[]{1,2,3,4};

     //append items to list:
     for(int item : items) {
        results = appendNode(ROOT, LAST, item);
        ROOT = results[0];
        LAST = results[1];
     }

     System.out.println("Compute the sum of the nodes:");
     System.out.println("=> list items:");
```

```
        int sum = 0;
        Node CURR = ROOT;
        while(CURR != null) {
          System.out.println("Node: "+CURR.data);
          sum += CURR.data;
          CURR = CURR.next;
        }

        System.out.println("Sum of nodes: "+sum);
    }
}
```

Listing 6.1 defines a `Node` class as before, followed by the method `appendNode()` thatcontains the logic for initializing a singly linked list and also for appending nodes to that list.

Launch the code in Listing 6.1 from the command line and you will see the following output:

```
list of numbers: [1 2 3 4]
=> list items:
Node: 1
Node: 2
Node: 3
Node: 4
Sum of nodes: 10
```

TASK: ADDING NUMBERS IN A LINKED LIST (2)

Listing 6.2 displays the contents of `SumSLNodes.java` that illustrates how to add the numbers in a linked list.

LISTING 6.2: SumSLNodes2.java

```
public class SumSLNodes2
{
   static class Node {
     int data;
     Node next = null;
     public Node(int data) {
       this.data = data;
       this.next = null;
     }
   }

   public static Node[] appendNode(Node ROOT, Node LAST,
   int item)
   {
     Node[] results = new Node[2];

     if(ROOT == null) {
```

```
        ROOT = new Node(item);
        //System.out.println("1Node: "+ROOT.data);
      } else {
        if(ROOT.next == null) {
          Node NEWNODE = new Node(item);
          LAST = NEWNODE;
          ROOT.next = LAST;
          //System.out.println("2Node: "+NEWNODE.data);
        } else {
          Node NEWNODE = new Node(item);
          LAST.next = NEWNODE;
          LAST = NEWNODE;
          //System.out.println("3Node: "+NEWNODE.data);
        }
      }

      //return ROOT, LAST
      results[0] = ROOT;
      results[1] = LAST;
      return results;
    }

    public static void main(String[] args)
    {
       Node ROOT  = null;
       Node LAST  = null;
       Node[] results = new Node[2];
       int[] items = new int[]{1,2,3,4};

       System.out.println("=> list items:");
       for(int num : items) {
         System.out.print(num+" ");
       }
       System.out.println("\n");

       //append items to list:
       for(int item : items) {
          results = appendNode(ROOT, LAST, item);
          ROOT = results[0];
          LAST = results[1];
       }

       // reconstruct the original number:
       // NB: [1,2,3,4] => 4,321

       int sum = 0, pow = 0, base = 10, term = 0;

       Node CURR = ROOT;
       while(CURR != null) {
         term = (int)java.lang.Math.pow(base,pow);
         term *= CURR.data;
```

```
            sum += term;
            pow += 1;
            CURR = CURR.next;
        }
        System.out.println("Reconstructed number: "+sum);
    }
}
```

Listing 6.2 defines a `Node` class as before, followed by the function `appendNode()` that contains the logic for initializing a singly linked list and also for appending nodes to that list.

Launch the code in Listing 6.2 from the command line and you will see the following output:

```
list of digits:   [1 2 3 4]
Original number:  4321
```

TASK: ADDING NUMBERS IN A LINKED LIST (3)

Listing 6.3 displays the contents of `SumSLNodes3.java` that illustrates how to add the numbers in a linked list.

LISTING 6.3: SumSLNodes3.java

```
public class SumSLNodes3
{
   static class Node {
     int data;
     Node next = null;
     public Node(int data) {
       this.data = data;
       this.next = null;
     }
   }

   public static Node[] appendNode(Node ROOT, Node LAST,
   int item)
   {
     Node[] results = new Node[2];

     if(ROOT == null) {
       ROOT = new Node(item);
       //System.out.println("1Node: "+ROOT.data);
     } else {
       if(ROOT.next == null) {
         Node NEWNODE = new Node(item);
         LAST = NEWNODE;
         ROOT.next = LAST;
         //System.out.println("2Node: "+NEWNODE.data);
       } else {
```

```
            Node NEWNODE = new Node(item);
            LAST.next = NEWNODE;
            LAST = NEWNODE;
            //System.out.println("3Node: "+NEWNODE.data);
         }
      }

      //return ROOT, LAST
      results[0] = ROOT;
      results[1] = LAST;
      return results;
   }

   public static void main(String[] args)
   {
      Node ROOT  = null;
      Node LAST  = null;
      Node[] results = new Node[2];
      int[] items = new int[]{1,2,3,4};

      System.out.println("=> list items:");
      for(int num : items) {
        System.out.print(num+" ");
      }
      System.out.println("\n");

      //append items to list:
      for(int item : items) {
         results = appendNode(ROOT, LAST, item);
         ROOT = results[0];
         LAST = results[1];
      }

      // reconstruct the reversed number:
      // NB: [1,2,3,4] => 1,234

      int sum = 0, pow = 0, base = 10;

      Node CURR = ROOT;
      while(CURR != null) {
        System.out.println("item: "+CURR.data+" sum: "+sum);
        sum = sum*base+ CURR.data;
        pow += 1;
        CURR = CURR.next;
      }
      System.out.println("Reconstructed number: "+sum);
   }
}
```

Listing 6.3 defines a `Node` class as before, followed by the method `appendNnode()` that contains the logic for initializing a singly linked list and also for appending nodes to that list. Launch the code in Listing 6.3 from the command line and you will see the following output:

```
=> list items:
1 2 3 4

item: 1 sum: 0
item: 2 sum: 1
item: 3 sum: 12
item: 4 sum: 123
Reconstructed number: 1234
```

TASK: DISPLAY THE FIRST K NODES

Listing 6.4 displays the contents of `FirstKNodes.java` that illustrates how to display the first k nodes in a linked list.

LISTING 6.4: FirstKNodes.java

```
public class FirstKNodes
{
   static class Node {
     String data = "";
     Node next = null;
     public Node(String data) {
       this.data = data;
       this.next = null;
     }
   }

   public static Node[] appendNode(Node ROOT, Node LAST,
   String item)
   {
     Node[] results = new Node[2];

     if(ROOT == null) {
       ROOT = new Node(item);
       //System.out.println("1Node: "+ROOT.data);
     } else {
       if(ROOT.next == null) {
         Node NEWNODE = new Node(item);
         LAST = NEWNODE;
         ROOT.next = LAST;
         //System.out.println("2Node: "+NEWNODE.data);
       } else {
         Node NEWNODE = new Node(item);
         LAST.next = NEWNODE;
         LAST = NEWNODE;
         //System.out.println("3Node: "+NEWNODE.data);
       }
```

```
    }

    //return ROOT, LAST
    results[0] = ROOT;
    results[1] = LAST;
    return results;
  }

  public static void firstKNodes(Node ROOT, int num)
  {
    int count = 0;
    Node CURR = ROOT;
    System.out.println("=> Display first "+num+" nodes");

    while (CURR != null) {
      count += 1;
      System.out.println("Node "+count+" data: "+CURR.data);
      CURR = CURR.next;

      if(count >= num)
        break;
    }
  }

  public static void main(String[] args)
  {
     Node[] results = new Node[2];
     Node ROOT  = null;
     Node LAST  = null;

     //append items to list:
     String[] items = new String[]
       {"Stan", "Steve", "Sally",
       "Alex","George","Fred","Bob"};

     for(String item : items) {
        results = appendNode(ROOT, LAST, item);
        ROOT = results[0];
        LAST = results[1];
     }

     System.out.println("=> Initial list:");
     for(String item : items) {
       System.out.println(item);
     }

     System.out.println();
     int node_count = 3;
     System.out.println("First "+node_count+" nodes:");
```

```
        firstKNodes(ROOT, node_count);
    }
}
```

Listing 6.4 defines a `Node` class as before, followed by the function `appendNode()` that contains the logic for initializing a singly linked list and also for appending nodes to that list.

Launch the code in Listing 6.4 from the command line and you will see the following output:

```
initial list:
=> Initial list:
Stan
Steve
Sally
Alex
George
Fred
Bob

First 3 nodes:
=> Display first 3 nodes
Node 1 data: Stan
Node 2 data: Steve
Node 3 data: Sally
```

TASK: DISPLAY THE LAST K NODES

Listing 6.5 displays the contents of `LastKNodes.java` that illustrates how to display the last k nodes of a linked list.

LISTING 6.5: LastKNodes.java

```
public class FirstKNodes
{
   static class Node {
     String data = "";
     Node next = null;
     public Node(String data) {
       this.data = data;
       this.next = null;
     }
   }

   public static Node[] appendNode(Node ROOT, Node LAST,
   String item)
   {
     Node[] results = new Node[2];

     if(ROOT == null) {
       ROOT = new Node(item);
```

```
      //System.out.println("1Node: "+ROOT.data);
    } else {
      if(ROOT.next == null) {
        Node NEWNODE = new Node(item);
        LAST = NEWNODE;
        ROOT.next = LAST;
        //System.out.println("2Node: "+NEWNODE.data);
      } else {
        Node NEWNODE = new Node(item);
        LAST.next = NEWNODE;
        LAST = NEWNODE;
        //System.out.println("3Node: "+NEWNODE.data);
      }
    }

    //return ROOT, LAST
    results[0] = ROOT;
    results[1] = LAST;
    return results;
  }

  public static int countNodes(Node ROOT)
  {
    int count = 0;
    Node CURR = ROOT;
    while (CURR != null) {
      count += 1;
      CURR = CURR.next;
    }
    return count;
  }

  public static Node skipNodes(Node ROOT, int skip_count)
{
    int count = 0;
    Node CURR = ROOT;
    while (CURR != null) {
      count += 1;
      CURR = CURR.next;
      if(count >= skip_count)
        break;
    }
    return CURR;
  }

  public static void lastKNodes(Node ROOT, int node_count,
  int num)
  {
    int count = 0;
    node_count = count_nodes(ROOT);
    Node START_NODE = skipNodes(ROOT,node_count-num);
```

```
      Node CURR = START_NODE;
      while (CURR != null) {
        count += 1;
        System.out.println("Node "+count+" data: "+CURR.data);
        CURR = CURR.next;

        if(count >= num)
          break;
      }
    }

    public static void main(String[] args)
    {
       Node[] results = new Node[2];
       Node ROOT  = null;
       Node LAST  = null;

       //append items to list:
       String[] items = new String[]
          {"Stan", "Steve", "Sally",
          "Alex","George","Fred","Bob"};

       for(String item : items) {
          results = appendNode(ROOT, LAST, item);
          ROOT = results[0];
          LAST = results[1];
       }

       System.out.println("=> Initial list:");
       for(String item : items) {
         System.out.println(item);
       }

       System.out.println();

       int list_length = count_nodes(ROOT);
       int node_count = 3;
       System.out.println("Last "+node_count+" nodes:");
       lastKNodes(ROOT, list_length,node_count);
    }
}
```

Listing 6.5 defines a `Node` class as before, followed by the method `appendNode()` that contains the logic for initializing a singly linked list and also for appending nodes to that list.

Launch the code in Listing 6.5 from the command line and you will see the following output:

```
=> Initial list:
Stan
Steve
Sally
Alex
George
Fred
Bob

Last 3 nodes:
Node 1 data: George
Node 2 data: Fred
Node 3 data: Bob
```

REVERSE A SINGLY LINKED LIST VIA RECURSION

Listing 6.6 displays `ReverseSLList.java` that illustrates how to update a node in a linked list.

LISTING 6.6: ReverseSLList.java

```
public class ReverseSLList
{
   static class Node {
     String data = "";
     Node next = null;
     public Node(String data) {
       this.data = data;
       this.next = null;
     }
   }

   public static Node[] appendNode(Node ROOT, Node LAST,
   String item)
   {
     Node[] results = new Node[2];

     if(ROOT == null) {
       ROOT = new Node(item);
       //System.out.println("1Node: "+ROOT.data);
     } else {
       if(ROOT.next == null) {
         Node NEWNODE = new Node(item);
         LAST = NEWNODE;
         ROOT.next = LAST;
         //System.out.println("2Node: "+NEWNODE.data);
       } else {
         Node NEWNODE = new Node(item);
         LAST.next = NEWNODE;
         LAST = NEWNODE;
```

```
          //System.out.println("3Node: "+NEWNODE.data);
        }
      }

      //return ROOT, LAST
      results[0] = ROOT;
      results[1] = LAST;
      return results;
    }

    public static String reverseList(Node node, String
    reversed_list)
    {
      if(node == null) {
        return reversed_list;
      } else {
        reversed_list = node.data + " "+ reversed_list;
        return reverseList(node.next, reversed_list);
      }
    }

    public static void main(String[] args)
    {
       Node ROOT = null;
       Node LAST = null;
       Node[] results = new Node[2];

       //append items to list:
       String[] items = new String[]
         {"Stan", "Steve", "Sally",
         "Alex","George","Fred","Bob"};

       for(String item : items) {
          results = appendNode(ROOT, LAST, item);
          ROOT = results[0];
          LAST = results[1];
       }

       System.out.println("=> Initial list:");
       for(String item : items) {
         System.out.print(item+" ");
       }
       System.out.println("\n");

       System.out.println("Reversed list:");
       String rev_list = "";
       String reversed = reverseList(ROOT, rev_list);
       System.out.println(reversed);
    }
}
```

Listing 6.6 defines a Node class as before, followed by the `Java` method `appendNode()` that contains the logic for initializing a singly linked list and also for appending nodes to that list.

Launch the code in Listing 6.6 from the command line and you will see the following output:

```
=> Initial list:
Stan Steve Sally Alex George Fred Bob

Reversed list:
Bob Fred George Alex Sally Steve Stan
```

TASK: REMOVE DUPLICATES

Listing 6.7 displays the contents of `RemoveDuplicates.java` that illustrates how to remove duplicate nodes in a linked list.

LISTING 6.7: RemoveDuplicates.java

```
public class RemoveDuplicates
{
   static class Node {
     String data = "";
     Node next = null;
     public Node(String data) {
       this.data = data;
       this.next = null;
     }
   }

   public static Node[] appendNode(Node ROOT, Node LAST,
   String item)
   {
      if(ROOT == null) {
        ROOT = new Node(item);
        //System.out.println("1Node:", ROOT.data)
      } else {
        if(ROOT.next == null) {
          Node NEWNODE = new Node(item);
          LAST = NEWNODE;
          ROOT.next = LAST;
          //System.out.println("2Node: "+NEWNODE.data)
        } else {
          Node NEWNODE = new Node(item);
          LAST.next = NEWNODE;
          LAST = NEWNODE;
          //System.out.println("3Node: "+NEWNODE.data);
        }
      }
```

```
    Node[] results = new Node[2];
    results[0] = ROOT;
    results[1] = LAST;
    return results;
  }

  public static Node deleteDuplicates(Node ROOT)
  {
    Node PREV = ROOT;
    Node CURR = ROOT;
    Boolean Found = false;

    System.out.println("=> searching for duplicates");
    int duplicate = 0;
    while (CURR != null) {
      Node SEEK = CURR;
      while (SEEK.next != null) {
        if(SEEK.next.data == CURR.data) {
          duplicate += 1;
          System.out.println("=> Duplicate node #"+
                duplicate+" with value: "+CURR.data);
          SEEK.next = SEEK.next.next;
        } else {
          SEEK = SEEK.next;
        }
      }
      CURR = CURR.next;
    }
    return ROOT;
  }

  public static void displayItems(Node ROOT)
  {
   System.out.println("=> list items:");
   Node CURR = ROOT;
   while(CURR != null) {
     System.out.println("Node: "+CURR.data);
     CURR = CURR.next;
   }
   System.out.println();
  }

  public static void main(String[] args)
  {
    Node[] results = new Node[2];
    Node ROOT  = null;
    Node LAST  = null;

    // append items to list:
    String[] items = new String[]
        {"Stan", "Steve", "Stan", "George","Stan"};
```

```
        for(String item : items) {
          results = appendNode(ROOT, LAST, item);
          ROOT = results[0];
          LAST = results[1];
        }

        displayItems(ROOT);

        String[] items2 = new String[]
          {"Stan", "Alex", "Sally", "Steve", "George"};
        for(String item2 : items2) {
          ROOT = deleteDuplicates(ROOT);
          displayItems(ROOT);
        }

        String[] items3 = new String[]
            {"Stan", "Steve", "Stan", "George","Stan"};
        System.out.println("original:");
        System.out.println(items3);

        System.out.println("unique:");
        displayItems(ROOT);
    }
}
```

Listing 6.7 defines a `Node` class, which is the same as earlier code samples in this chapter, followed by the function `appendNode()` that contains the logic for initializing a singly linked list and also for appending nodes to that list.

Launch the code in Listing 6.7 from the command line and you will see the following output:

```
=> list items:
Node: Stan
Node: Steve
Node: Stan
Node: George
Node: Stan

=> searching for duplicates
=> Duplicate node #1 with value: Stan
=> Duplicate node #2 with value: Stan
=> list items:
Node: Stan
Node: Steve
Node: George

=> searching for duplicates
=> list items:
Node: Stan
Node: Steve
Node: George
```

```
=> searching for duplicates
=> list items:
Node: Stan
Node: Steve
Node: George

=> searching for duplicates
=> list items:
Node: Stan
Node: Steve
Node: George

=> searching for duplicates
=> list items:
Node: Stan
Node: Steve
Node: George

original:
[Ljava.lang.String;@6d06d69c
unique:
=> list items:
Node: Stan
Node: Steve
Node: George
```

TASK: CONCATENATE TWO LISTS

Listing 6.8 displays the contents of `AppendSLLists.java` that illustrates how to split a linked list into two lists.

LISTING 6.8: AppendSLLists.java

```
public class AppendSLLists
{
   static class Node {
     int data;
     Node next = null;
     public Node(int data) {
       this.data = data;
       this.next = null;
     }
   }

   public static Node[] appendNode(Node ROOT, Node LAST,
   int item)
   {
     Node[] results = new Node[2];

     if(ROOT == null) {
```

```
        ROOT = new Node(item);
        //System.out.println("1Node: "+ROOT.data);
      } else {
        if(ROOT.next == null) {
          Node NEWNODE = new Node(item);
          LAST = NEWNODE;
          ROOT.next = LAST;
          //System.out.println("2Node: "+NEWNODE.data);
        } else {
          Node NEWNODE = new Node(item);
          LAST.next = NEWNODE;
          LAST = NEWNODE;
          //System.out.println("3Node: "+NEWNODE.data);
        }
      }

      //return ROOT, LAST
      results[0] = ROOT;
      results[1] = LAST;
      return results;
    }

    public static void displayItems(Node ROOT)
    {
      Node CURR = ROOT;
      while(CURR != null) {
        System.out.println("Node: "+CURR.data);
        CURR = CURR.next;
      }
      System.out.println();
    }

    public static void main(String[] args)
    {
       Node[] results = new Node[2];

       //Node ROOT  = null;
       //Node LAST  = null;
       Node node2 = null;

       int index = 2, count = 0;

       // append items to list1:
       Node ROOT1 = null;
       Node LAST1 = null;
       //items1 = np.array{300, 50, 30, 80, 100, 200}
       int[] items1 = new int[]{300, 50, 30};

       for(int item : items1) {
         results = appendNode(ROOT1, LAST1, item);
```

```
            ROOT1 = results[0];
            LAST1 = results[1];
            if(count == index)
              node2 = LAST1;
            count += 1;
          }

          System.out.println("FIRST LIST:");
          displayItems(ROOT1);

          Node ROOT2 = null;
          Node LAST2 = null;
          //append second set of items to list:
          int[] items2 = new int[]{80, 100, 200};
          for(int item : items2) {
            results = appendNode(ROOT2, LAST2, item);
            ROOT2 = results[0];
            LAST2 = results[1];
            if(count == index)
              node2 = LAST2;
            count += 1;
          }

          System.out.println("SECOND LIST:");
          displayItems(ROOT2);

          // concatenate the two lists:
          System.out.println("COMBINED LIST:");
          LAST1.next = ROOT2;
          displayItems(ROOT1);
      }
}
```

Listing 6.8 starts by defining the class Node whose contents you have seen in previous examples in this chapter. The remaining code consists of four static methods: appendNode(), displayItems, and main() for appending a node, displaying all items, and the main() method of this Java class, respectively.

The appendNode() and displayItems() methods also contain the same code that you have seen in previous code samples in this chapter. The main() method consists of two blocks of code. The first block of code constructs (and then displays) the first linked list based on the elements in the integer array items1, and the second block of code constructs (and then displays) the second linked list based on the elements in the integer array items2. Notice that both blocks of code invoke the appendNode() method in order to construct the two lists. Launch the code in Listing 6.8 and you will see the following output:

```
FIRST LIST:
Node: 300
Node: 50
Node: 30
```

```
SECOND LIST:
Node: 80
Node: 100
Node: 200

COMBINED LIST:
Node: 300
Node: 50
Node: 30
Node: 80
Node: 100
Node: 200
```

TASK: MERGE TWO ORDERED LINKED LISTS

Listing 6.9 displays the contents of `MergeSLLists.java` that illustrates how to merge two linked lists.

LISTING 6.9: MergeSLLists.java

```
public class MergeSLLists
{
   static class Node {
     int data;
     Node next = null;
     public Node(int data) {
       this.data = data;
       this.next = null;
     }
   }

   public static Node[] appendNode(Node ROOT, Node LAST,
   int item)
   {
     Node[] results = new Node[2];

     if(ROOT == null) {
       ROOT = new Node(item);
       //System.out.println("1Node: "+ROOT.data);
     } else {
       if(ROOT.next == null) {
         Node NEWNODE = new Node(item);
         LAST = NEWNODE;
         ROOT.next = LAST;
         //System.out.println("2Node: "+NEWNODE.data);
       } else {
         Node NEWNODE = new Node(item);
         LAST.next = NEWNODE;
         LAST = NEWNODE;
         //System.out.println("3Node: "+NEWNODE.data);
```

```
      }
    }

    //return ROOT, LAST
    results[0] = ROOT;
    results[1] = LAST;
    return results;
  }

  public static void displayItems(Node ROOT)
  {
    Node CURR = ROOT;
    while(CURR != null) {
      System.out.println("Node: "+CURR.data);
      CURR = CURR.next;
    }
    System.out.println();
  }

  public static void main(String[] args)
  {
     Node[] results = new Node[2];
     Node node2 = null;

     int index = 2, count = 0;

     // create the first list:
     Node ROOT1 = null;
     Node LAST1 = null;
     int[] items1 = new int[]{30, 50, 300};

     for(int item : items1) {
       results = appendNode(ROOT1, LAST1, item);
       ROOT1 = results[0];
       LAST1 = results[1];
       if(count == index)
         node2 = LAST1;
       count += 1;
     }

     System.out.println("FIRST LIST:");
     displayItems(ROOT1);

     Node ROOT2 = null;
     Node LAST2 = null;

     // create a second list:
     int[] items2 = new int[]{80, 100, 200};
     for(int item : items2) {
       results = appendNode(ROOT2, LAST2, item);
       ROOT2 = results[0];
```

```
    LAST2 = results[1];
    if(count == index)
      node2 = LAST2;
    count += 1;
  }

  System.out.println("SECOND LIST:");
  displayItems(ROOT2);

  Node CURR1 = ROOT1;
  LAST1 = ROOT1;
  Node CURR2 = ROOT2;
  LAST2 = ROOT2;
  Node ROOT3 = null;
  Node LAST3 = null;

  while(CURR1 != null && CURR2 != null) {
    //System.out.println("curr1.data: "+CURR1.data);
    //System.out.println("curr2.data: "+CURR2.data);

    if(CURR1.data < CURR2.data) {
      results = appendNode(ROOT3, LAST3, CURR1.data);
      ROOT3 = results[0];
      LAST3 = results[1];

      //System.out.println("adding curr1.data: "+CURR1.
      data);
      CURR1 = CURR1.next;
    } else {
      results = appendNode(ROOT3, LAST3, CURR2.data);
      ROOT3 = results[0];
      LAST3 = results[1];

      //System.out.println("adding curr2.data: "+CURR2.
      data);
      CURR2 = CURR2.next;
    }
  }

  // append any remaining elements of items1:
  if(CURR1 != null) {
    while(CURR1 != null) {
      //System.out.println("MORE curr1.data: "+CURR1.
      data);
      results = appendNode(ROOT3, LAST3, CURR1.data);
      ROOT3 = results[0];
      LAST3 = results[1];

      CURR1 = CURR1.next;
    }
  }
```

```
        // append any remaining elements of items2:
        if(CURR2 != null) {
          while(CURR2 != null) {
            System.out.println("MORE curr2.data: "+CURR2.
            data);
            results = appendNode(ROOT3, LAST3, CURR2.data);
            ROOT3 = results[0];
            LAST3 = results[1];
            CURR2 = CURR2.next;
          }
        }

        System.out.println("MERGED LIST:");
        displayItems(ROOT3);
    }
}
```

Listing 6.9 starts by defining the class Node whose contents you have seen in previous examples in this chapter. The remaining code consists of four static methods: appendNode(), displayItems(), and main() for appending a node, displaying all items, and the main() method of this Java class, respectively.

The appendNode() and displayItems() methods also contain the same code that you have seen in previous code samples in this chapter. The main() method consists of two blocks of code. The first block of code constructs (and then displays) the first linked list based on the elements in the integer array items1, and the second block of code constructs (and then displays) the second linked list based on the elements in the integer array items2. Notice that both blocks of code invoke the appendNode() method in order to construct the two lists.

The next portion of the main() method contains a loop that constructs a third list that is based on the contents of the previous pair of lists, and the result is an ordered list of numbers (i.e., from smallest to largest). *This code works correctly because the initial pair of lists are ordered in numerically increasing order*. During each iteration, the following conditional logic is performed, where item1 and item2 are the current elements of list1 and list2, respectively:

If item1 < item2, append item1 to list3; otherwise append item2 to list3.

The loop finishes executing when the last item of either list is processed. Of course, there might be additional unprocessed elements in either list1 or list2 (but not both), so two additional loops are required: one loop appends the remaining items in list1 (if any) to list3, and another loop appends the remaining items in list2 (if any) to list3. Launch the code in Listing 6.8 and you will see the following output:

```
FIRST LIST:
Node: 30
Node: 50
Node: 300
```

```
SECOND LIST:
Node: 80
Node: 100
Node: 200

MERGED LIST:
Node: 30
Node: 50
Node: 80
Node: 100
Node: 200
Node: 300
```

TASK: SPLIT AN ORDERED LIST INTO TWO LISTS

There are several ways to perform this task. One approach is to iterate through a given list and dynamically create a list of smaller items as well as a list of larger items in the loop. However, the logic is more complex, and therefore more error prone.

A simpler approach involves appending the smaller items to a list and then appending the remaining items to a larger list, and then return the two lists. At this point you can invoke the `append()` function to create two linked lists.

Listing 6.10 displays the contents of `SplitSLLists.java` that illustrates how to split a linked list into two lists.

LISTING 6.10: SplitSLLists.java

```
import java.util.ArrayList;

public class SplitSLLists
{
   static class Node {
     int data;
     Node next = null;

     public Node(int data) {
       this.data = data;
       this.next = null;
     }
   }

   public static Node[] appendNode(Node ROOT, Node LAST,
   int item)
   {
     Node[] results = new Node[2];

     if(ROOT == null) {
       ROOT = new Node(item);
```

```
      //System.out.println("1Node: "+ROOT.data);
    } else {
      if(ROOT.next == null) {
        Node NEWNODE = new Node(item);
        LAST = NEWNODE;
        ROOT.next = LAST;
        //System.out.println("2Node: "+NEWNODE.data);
      } else {
        Node NEWNODE = new Node(item);
        LAST.next = NEWNODE;
        LAST = NEWNODE;
        //System.out.println("3Node: "+NEWNODE.data);
      }
    }

    //return ROOT, LAST
    results[0] = ROOT;
    results[1] = LAST;
    return results;
  }

  public static void deleteNode(Node node)
  {
    Boolean Found = false;

    if(node != null) {
      if(node.next != null) {
        Found = true;
        System.out.println("curr node: "+node.data);
        node.data = node.next.data;
        node.next = node.next.next;
        System.out.println("new  node: "+node.data);
      }
    }

    if(Found == false)
     System.out.println("* Item "+node.data+" not in list *");
  }

  public static ArrayList[] splitList(Node ROOT, int value)
  {
    ArrayList[] results = new ArrayList[2];

    Node node = ROOT;
    ArrayList smaller = new ArrayList();
    ArrayList larger  = new ArrayList();

    System.out.println("Comparison value: "+value);
    while(node != null) {
      if(node.data < value) {
        //System.out.println("LESS curr node: "+node.data);
        smaller.add(node.data);
```

```
      } else {
        //System.out.println("GREATER curr node: "+node.data);
        larger.add(node.data);
      }
      node = node.next;
    }
    results[0] = smaller;
    results[1] = larger;

    return results;
  }

  public static void displayItems(Node ROOT)
  {
    Node CURR = ROOT;
    while(CURR != null) {
      System.out.println("Node: "+CURR.data);
      CURR = CURR.next;
    }
    System.out.println();
  }

  public static void main(String[] args)
  {
     Node[] results1 = new Node[2];
     ArrayList[] results2 = new ArrayList[2];
     Node node2 = null;

     int index = 2, count = 0;

     // append items to list1:
     Node ROOT  = null;
     Node LAST  = null;
     int[] items1 = new int[]{300, 50, 30};

     for(int item : items1) {
       results1 = appendNode(ROOT, LAST, item);
       ROOT = results1[0];
       LAST = results1[1];

       if(count == index)
         node2 = LAST;
       count += 1;
     }

     System.out.println("=> FIRST LIST:");
     displayItems(ROOT);

     int value = 70; //  node2.data;
     results2 = splitList(ROOT, value);
     ArrayList smaller = results2[0];
     ArrayList larger  = results2[1];
```

```
        System.out.println("smaller list: "+smaller);
        System.out.println("larger  list: "+larger);
    }
}
```

Listing 6.10 starts by defining the class Node whose contents you have seen in previous examples in this chapter. The remaining code consists of four static methods: appendNode(), splitList(), displayItems(), and main() for appending a node, splitting a list into two lists, displaying all items, and the main() method of this Java class, respectively.

The appendNode() and displayItems() methods also contain the same code that you have seen in previous code samples in this chapter. The splitList() splits a given list into two distinct lists. Specifically, the first list consists of the numbers that are less than a comparison value, and the second list consists of the remaining elements. Since the initial list is ordered, the code for this method involves a loop that performs a comparison of each list element with the comparison value.

The main() method starts by constructing (and then displays) the initial linked list from the elements in the integer array items1. The next portion of the main() method invokes the splitList() method that constricts two sublists (one of which can be empty) and returns the resulting lists. The final portion of the main() method displays the contents of the two sublists. Launch the code in Listing 6.10 and you will see the following output:

```
=> list items:
Node: 300
Node: 50
Node: 30

GREATER curr node: 300
GREATER curr node: 50
GREATER curr node: 30
smaller list: []
larger  list: [300, 50, 30]
```

TASK: REMOVE A GIVEN NODE FROM A LIST

This task starts by constructing a linked list of items, after which an intermediate node is removed from the list.

Listing 6.11 displays the contents of `DeleteSLNode2.java` that illustrates how to delete a node in a singly linked list.

LISTING 6.11: DeleteSLNode2.java

```
public class DeleteSLNode2
{
   static class Node {
```

```
    String data = "";
    Node next = null;
    public Node(String data) {
      this.data = data;
      this.next = null;
    }
  }

  public static Node[] appendNode(Node ROOT, Node LAST,
  String item)
  {
    Node[] results = new Node[2];

    if(ROOT == null) {
      ROOT = new Node(item);
      //System.out.println("1Node: "+ROOT.data);
    } else {
      if(ROOT.next == null) {
        Node NEWNODE = new Node(item);
        LAST = NEWNODE;
        ROOT.next = LAST;
        //System.out.println("2Node: "+NEWNODE.data);
      } else {
        Node NEWNODE = new Node(item);
        LAST.next = NEWNODE;
        LAST = NEWNODE;
        //System.out.println("3Node: "+NEWNODE.data);
      }
    }

    results[0] = ROOT;
    results[1] = LAST;
    return results;
  }

  public static void deleteNode(Node node)
  {
    Boolean Found = false;

    System.out.println("Delete node: "+node.data);
    if(node != null) {
      if(node.next != null) {
        Found = true;
        //System.out.println("curr node: "+node.data);
        node.data = node.next.data;
        node.next = node.next.next;
        //System.out.println("new  node: "+node.data);
      }
    }

    if(Found == false)
```

```
        System.out.println("* Item "+node.data+" not in list *");
    }

    public static void displayItems(Node ROOT)
    {
      System.out.println("=> list items:");
      Node CURR = ROOT;
      while(CURR != null) {
        System.out.println("Node: "+CURR.data);
        CURR = CURR.next;
      }
      System.out.println();
    }

    public static void main(String[] args)
    {
       Node[] results = new Node[2];
       Node ROOT  = null;
       Node LAST  = null;
       Node node2 = null;

       int index=2, count = 0;
       String[] items = new String[]
            {"Stan", "Steve", "Sally", "George","Alex"};

       for(String item : items) {
          results = appendNode(ROOT, LAST, item);
          ROOT = results[0];
          LAST = results[1];
          if(count == index)
            node2 = LAST;
          count += 1;
       }

       displayItems(ROOT);
       deleteNode(node2);
       displayItems(ROOT);
    }
}
```

Listing 6.11 starts by defining the class Node whose contents you have seen in previous examples in this chapter. The remaining code consists of the static methods appendNode(), deleteNode(), displayItems(), and main() for appending a node, deleting a node, displaying all items, and the main() method of this Java class, respectively.

The appendNode() and displayItems() methods also contain the same code that you have seen in previous code samples in this chapter. The delete() method deletes a given node by iterating through the given list. If the node is found in the list, then the "next" pointer of its predecessor is updated so that it points to the successor of the node that will be deleted.

The main() method starts by constructing (and then displaying) the initial linked list from the elements in the items array. The next portion of the main() method invokes the displayItems() method, the deleteNode() method, and the displayItems() method. Launch the code in Listing 6.10 and you will see the following output:

```
=> list items:
Node: Stan
Node: Steve
Node: Sally
Node: George
Node: Alex

Delete node: Sally
=> list items:
Node: Stan
Node: Steve
Node: George
Node: Alex
```

TASK: FIND THE MIDDLE ELEMENT IN A LIST

One solution involves counting the number of elements in the list and then finding the middle element. However, this task has the following constraints:

- Counting the number of elements is not allowed
- No additional data structure can be used
- No element in the list can be modified or marked
- Lists of even length can have two middle elements

This task belongs to a set of tasks that use the same technique: one variable iterates sequentially through a list and a second variable iterates twice as quickly through the same list. In fact, this technique is used to determine whether or not a list contains a loop (which is discussed in the next section).

Listing 6.12 displays the contents of `MiddleSLNode.py` that illustrates how to determine the middle element in. Singly linked list.

LISTING 6.12: MiddleSLNode.java

```
public class MiddleSLNode
{
   static class Node {
     String data = "";
     Node next = null;

     public Node(String data) {
       this.data = data;
       this.next = null;
```

```
    }
  }

  public static Node[] appendNode(Node ROOT, Node LAST,
  String item)
  {
    Node[] results = new Node[2];

    if(ROOT == null) {
      ROOT = new Node(item);
      //System.out.println("1Node: "+ROOT.data);
    } else {
      if(ROOT.next == null) {
        Node NEWNODE = new Node(item);
        LAST = NEWNODE;
        ROOT.next = LAST;
        //System.out.println("2Node: "+NEWNODE.data);
      } else {
        Node NEWNODE = new Node(item);
        LAST.next = NEWNODE;
        LAST = NEWNODE;
        //System.out.println("3Node: "+NEWNODE.data);
      }
    }

    results[0] = ROOT;
    results[1] = LAST;
    return results;
  }

  public static Node[] findMiddle(Node ROOT)
  {
    Node[] results = new Node[2];
    Node empty  = new Node("empty");
    Node even   = new Node("even");
    Node odd    = new Node("odd");
    Node result = new Node("");
    Node CURR   = ROOT;
    Node SEEK   = ROOT;

    int count = 0;
    if(ROOT == null) {
      results[0] = empty;
      results[1] = even;
      return results;
    } else if (ROOT.next == null) {
      count += 1;
      results[0] = ROOT;
      results[1] = odd;
      return results;
    }
```

```
    count += 1;

    while (SEEK != null) {
      //System.out.println("=> SEEK: "+SEEK.data);
      //System.out.println("=> CURR: "+CURR.data);
      //System.out.println("count:   "+count);

      if(SEEK.next == null) {
        //System.out.println("1break: null node");
        break;
      } else if(SEEK.next.next == null) {
        //System.out.println("2break: null node");
        count += 1;
        break;
      } else {
        SEEK = SEEK.next.next;
        CURR = CURR.next;
      }
      //count += 1;
    }

    if(count % 2 == 0) {
      result = even;
      System.out.println("=> setting even node");
    } else {
      System.out.println("=> setting odd node");
      result = odd;
    }

    results[0] = CURR;
    results[1] = result;
    return results;
  }

  public static void main(String[] args)
  {
     Node[] results = new Node[2];
     Node ROOT  = null;
     Node LAST  = null;

     //lists of even length and odd length:
     String[] items = new String[]
               {"Stan", "Steve", "Sally", "Alex"};

   //String[] items = new String[]
   //          {"Stan", "Steve", "Sally", "Alex","Dave"};

     for(String item : items) {
        results = appendNode(ROOT, LAST, item);
        ROOT = results[0];
```

```
            LAST = results[1];
        }

        System.out.println();
        System.out.println("=> list items:");

        Node CURR = ROOT;
        while(CURR != null) {
          System.out.println("Node:  "+CURR.data);
          CURR = CURR.next;
        }

        results = findMiddle(ROOT);
        if(results[1].data == "even")
         System.out.println("list has an even number of items");
        else
         System.out.println("list has an odd number of items");
    }
}
```

Listing 6.12 starts by defining the class Node whose contents you have seen in previous examples in this chapter. The remaining code consists of the static methods appendNode(), findMiddle(), and main() for appending a node, finding the middle node, and the main() method of this Java class, respectively.

The appendNode() and displayItems() methods also contain the same code that you have seen in previous code samples in this chapter. The findMiddle() method iterates through a given list to find the middle element: if the list has odd length, then there is a single element, otherwise there are two elements that are "tied" for the middle position. Notice that the loop iterates through the given list with the variables CURR and SEEK, both of which are node elements. Moreover, the CURR element iterates linearly, whereas the SEEK node uses a "double time" technique whereby the SEEK node advances twice as fast as CURR through the given list. Hence, when the SEEK node reaches the end of the list, the CURR node is positioned at the midpoint of the given list.

The main() method starts by constructing (and then displaying) the initial linked list from the elements in the items array. The next portion of the main() method invokes the findMiddle() method to determine the midpoint of the list. Note that the findMiddle() method returns the string "even" if the list has even length; otherwise, the list has odd length and in both cases an appropriate message is printed. Launch the code in Listing 6.12 and you will see the following output:

```
=> list items:
Node: Stan
Node: Steve
Node: Sally
Node: Alex
```

```
Node: Dave
=> setting odd node
list has an odd number of items
```

TASK: REVERSE A LINKED LIST

Listing 6.13 displays the contents of `ReverseSLList.java` that illustrates how to reverse the elements in a linked list.

LISTING 6.13: ReverseSLList.java

```
public class ReverseSLList
{
   static class Node {
     String data = "";
     Node next = null;
     public Node(String data) {
       this.data = data;
       this.next = null;
     }
   }

   public static Node[] appendNode(Node ROOT, Node LAST,
   String item)
   {
     Node[] results = new Node[2];

     if(ROOT == null) {
       ROOT = new Node(item);
       //System.out.println("1Node: "+ROOT.data);
     } else {
       if(ROOT.next == null) {
         Node NEWNODE = new Node(item);
         LAST = NEWNODE;
         ROOT.next = LAST;
         //System.out.println("2Node: "+NEWNODE.data);
       } else {
         Node NEWNODE = new Node(item);
         LAST.next = NEWNODE;
         LAST = NEWNODE;
         //System.out.println("3Node: "+NEWNODE.data);
       }
     }

     //return ROOT, LAST
     results[0] = ROOT;
     results[1] = LAST;
     return results;
   }
```

```
    public static String reverseList(Node node, String
    reversed_list)
    {
      if(node == null) {
        return reversed_list;
      } else {
        reversed_list = node.data + " "+ reversed_list;
        return reverseList(node.next, reversed_list);
      }
    }

    public static void main(String[] args)
    {
       Node ROOT = null;
       Node LAST = null;
       Node[] results = new Node[2];

       //append items to list:
       String[] items = new String[]
         {"Stan", "Steve", "Sally",
         "Alex","George","Fred","Bob"};

       for(String item : items) {
          results = appendNode(ROOT, LAST, item);
          ROOT = results[0];
          LAST = results[1];
       }

       System.out.println("=> Initial list:");
       for(String item : items) {
         System.out.print(item+" ");
       }
       System.out.println("\n");

       System.out.println("Reversed list:");
       String rev_list = "";
       String reversed = reverseList(ROOT, rev_list);
       System.out.println(reversed);
    }
}
```

The remaining code consists of the static methods appendNode(), reverseList(), and main() for appending a node, reversing the list, and the main() method of this Java class, respectively.

The appendNode() method contains the same code that you have seen in previous code samples in this chapter. The reverseList() method iterates through a given list to reverse its contents.

The main() method starts by constructing (and then displaying) the initial linked list from the elements in the items array. The next portion of the main()

method invokes the reverseList() method and then displays the contents of the reversed list. Launch the code in Listing 6.12 and you will see the following output:

```
=> Initial list:
Stan Steve Sally Alex George Fred Bob

Reversed list:
Bob Fred George Alex Sally Steve Stan
```

TASK: CHECK FOR PALINDROME IN A LINKED LIST

Listing 6.14 displays the contents of `PalindromeSLList.java` that illustrates how to determine whether or not a list contains a palindrome.

LISTING 6.14: PalindromeSLList.java

```
public class PalindromeSLList
{
   static class Node {
     String data = "";
     Node next = null;
     public Node(String data) {
       this.data = data;
       this.next = null;
     }
   }

   public static Node[] appendNode(Node ROOT, Node LAST,
   String item)
   {
     Node[] results = new Node[2];

     if(ROOT == null) {
       ROOT = new Node(item);
       //System.out.println("1Node: "+ROOT.data);
     } else {
       if(ROOT.next == null) {
         Node NEWNODE = new Node(item);
         LAST = NEWNODE;
         ROOT.next = LAST;
         //System.out.println("2Node: "+NEWNODE.data);
       } else {
         Node NEWNODE = new Node(item);
         LAST.next = NEWNODE;
         LAST = NEWNODE;
         //System.out.println("3Node: "+NEWNODE.data);
       }
     }
```

```
      results[0] = ROOT;
      results[1] = LAST;
      return results;
    }

    public static String reverseList(Node node, String
    reversed_list)
    {
      if(node == null) {
        return reversed_list;
      } else {
        reversed_list = node.data + " "+ reversed_list;
        return reverseList(node.next, reversed_list);
      }
    }

    public static void main(String[] args)
    {
       String[] items = new String[]{"a", "b", "c", "b", "d"};

       Node[] results = new Node[2];
       Node ROOT = null;
       Node LAST = null;

       for(String item : items) {
          results = appendNode(ROOT, LAST, item);
          ROOT = results[0];
          LAST = results[1];
       }

       String original = "";
       System.out.println("=> Original items:");
       Node CURR1 = ROOT;
       while(CURR1 != null) {
         original = CURR1.data+" ";
         CURR1 = CURR1.next;
       }
       original += "x";
       System.out.println(original);

       String rev_list = "";
       String reversed = reverseList(ROOT, rev_list);
       System.out.println("=> Reversed items:");
       System.out.println(reversed+"x");

       if(original.equals(rev_list))
         System.out.println("found a palindrome");
       else
         System.out.println("not a palindrome");
    }
}
```

Listing 6.14 defines a `Node` class as before, followed by the method `appendNode()` that contains the logic for initializing a singly linked list and also for appending nodes to that list.

Launch the code in Listing 6.14 from the command line and you will see the following output:

```
=> list of items:
a b c b a

=> New list of items:
a b c b a

found a palindrome
```

SUMMARY

This chapter started with code samples for displaying the first `k` nodes in a list as well as the last `k` nodes of a list. Then you learned how to display the contents of a list in reverse order and how to remove duplicates.

In addition, you saw how to concatenate and merge two linked lists, and how to split a single linked list. Then you learned how to remove the middle element in a list and how to determine whether or not a linked list contains a loop.

Finally, you learned how to calculate the sum of the elements in a linked list and how to check for palindromes in a linked list.

CHAPTER 7

QUEUES AND STACKS

This chapter introduces you to queues and stacks that were briefly introduced in Chapter 4 in the section pertaining to linear data structures.

The first part of this chapter explains the concept of a queue, along with `Java` code samples that show you how to perform various operations on a queue. Some of the code samples also contain built-in functions for queues, such as `isEmpty()`, `isFull()`, `push()`, and `dequeue()`.

The second part of this chapter explains the concept of a stack, along with `Java` code samples that show you how to perform various operations on a stack. In addition, you will see code samples for finding the largest and smallest elements in a stack and reversing the contents of a stack.

The final section contains three interesting tasks that illustrate the usefulness of a stack data structure. The first task determines whether or not a string consists of well-balanced round parentheses, square brackets, and curly braces. The second task parses an arithmetic expression that can perform addition, subtraction, multiplication, or division, as well as any combination of these four arithmetic operations. The third task converts infix notation to postfix notation.

WHAT IS A QUEUE?

A queue consists of a collection of objects that uses the FIFO (first-in-first-out) rule for inserting and removing items. By way of analogy, consider a toll booth: the first vehicle that arrives is the first vehicle to pay the necessary toll, and it's also the first vehicle to exit the tool booth. As another analogy, consider customers standing in a line (which in fact is a queue) in a bank: the person at the front of the queue is the first person to approach an available teller. The "back" of the queue is the person at the end of the line (i.e., the last person).

A queue has a maximum size `MAX` and a minimum size of 0. In fact, we can define a queue in terms of the following methods:

- `isEmpty()` returns True if the queue is empty
- `isFull()` returns True if the queue is full
- `queueSize()` returns the number of elements in the stack
- `add(item)` adds an element to the back of the queue if the queue is not full
- `dequeue()` removes the front element of the queue if the queue is not empty

In order to ensure that there is no underflow (empty) or overflow (full), we must always invoke `isEmpty()` before "popping" an item from the top of the front of a queue and always invoke `isFull()` before "pushing" (appending) an item as the last element of a queue.

Types of Queues

The following list various types of queues that can be created, most of which are extensions of a generic queue, followed by a brief description:

- queue
- circular queue
- dequeue
- priority queue

A *queue* is a linear list that supports deletion from one end and insertion at the other end. A queue is a FIFO, just like a line of people waiting to enter a movie theatre or a restaurant: the first person in line enters first, followed by the second person in line, and so forth. The term enqueue refers to adding an element to a queue, whereas dequeue refers to removing an element from a queue.

A *circular queue* is a linear list with the following constraint: the last element in the queue "points" to the first element in the queue. A circular queue is also called a *ring buffer*. By way of analogy, a conga line is a queue: if the person at the front of the queue is "connected" to the last person in the conga line, a *circular queue* has been created.

A *dequeue* is a linear list that is also a double ended queue in which insertions and deletions can be performed at *both* ends of the queue. In addition, there are two types of dequeues:

- Input restricted means that insertions occur only at one end.
- Output restricted means that deletions occur only at one end.

A *priority queue* is a queue that allows for removing and inserting items in any position of the queue. For example, the scheduler of the operating system of your desktop and laptop uses a priority queue to schedule programs for execution. Consequently, a higher priority task is executed before a lower priority task.

Moreover, after a priority queue is created, it's possible for a higher priority task to arrive in this scenario, that new and higher priority task is inserted into the appropriate location in the queue for task execution. In fact, Unix has the so-called `nice` command that you can launch from the command line in order to lower the execution priority of tasks. Perform an online search for more information regarding the queues discussed in this section.

Let's turn our attention to creating a basic queue, which is the topic of the next section.

CREATING A QUEUE IN JAVA

Listing 7.1 displays the contents of `MyQueue.java` that illustrates how to use a `Java ArrayList` as a data structure in order to perform various operations on a queue.

LISTING 7.1: MyQueue.java

```
import java.util.ArrayList;

public class MyQueue
{
   static int MAX = 4;
   static ArrayList myqueue = new ArrayList();

   public static Boolean isEmpty() {
     return myqueue.size() == 0;
   }

   public static Boolean isFull() {
     return myqueue.size() == MAX;
   }

   public static void dequeue() {
      if(myqueue.size() > 0) {
        int front = (int)myqueue.get(0);
        myqueue.remove(0);
        System.out.println("removed item : "+front);
      } else {
        System.out.println("* myqueue is empty *");
      }
   }

   public static void push(int item) {
      if(isFull() == false) {
        myqueue.add(item);
      } else {
        System.out.println("* myqueue is full *");
      }
   }

   public static void main(String[] args)
```

```
    {
       System.out.println("=> pushing values onto
       myqueue:");
       push(10);
       System.out.println("myqueue: "+myqueue);
       push(20);
       System.out.println("myqueue: "+myqueue);
       push(200);
       System.out.println("myqueue: "+myqueue);
       push(50);
       System.out.println("myqueue: "+myqueue);
       push(-123);
       System.out.println("myqueue: "+myqueue);
       System.out.println();

       System.out.println("=> dequeue values from myqueue: ");
       dequeue();
       System.out.println("myqueue: "+myqueue);
       dequeue();
       System.out.println("myqueue: "+myqueue);
       dequeue();
       System.out.println("myqueue: "+myqueue);
       dequeue();
       System.out.println("myqueue: "+myqueue);
       dequeue();
       System.out.println("myqueue: "+myqueue);
    }
}
```

Listing 7.1 starts by initializing `myqueue` as an empty list and assigning the value 4 to the variable `MAX`, which is the maximum number of elements that the queue can contain (obviously you can change this value).

The next portion of Listing 7.1 defines several functions: the `isEmpty` method that returns `True` if the length of `myqueue` is 0 (and false otherwise), followed by the function `isFull()` that returns `True` if the length of `myqueue` is `MAX` (and `False` otherwise).

The next portion of Listing 7.1 defines the method `dequeue` that invokes the `pop()` method in order to remove the front element of `myqueue`, provided that `myqueue` is not empty. Next, the function `push()` invokes the `append()` method in order to add a new element to the end of `myqueue`, provided that `myqueue` is not full.

The final portion of Listing 7.1 invokes the `push()` method to append various numbers to `myqueue`, followed by multiple invocations of the `dequeue()` method to remove elements from the front of the queue. Launch the code in Listing 7.1 and you will see the following output:

```
=> pushing values onto myqueue:
myqueue: [10]
myqueue: [10, 20]
```

```
myqueue: [10, 20, 200]
myqueue: [10, 20, 200, 50]
* myqueue is full *
myqueue: [10, 20, 200, 50]

=> dequeue values from myqueue:
removed item : 10
myqueue: [20, 200, 50]
removed item : 20
myqueue: [200, 50]
removed item : 200
myqueue: [50]
removed item : 50
myqueue: []
* myqueue is empty *
myqueue: []
```

Listing 7.2 displays the contents of `MyQueue2.java` that illustrates how to define a queue and perform various operations on the queue.

LISTING 7.2: MyQueue2.java

```
import java.util.ArrayList;

public class MyQueue2
{
   static int MAX = 4;
   static ArrayList myqueue = new ArrayList();

   public static Boolean isEmpty() {
     return myqueue.size() == 0;
   }

   public static Boolean isFull() {
     return myqueue.size() == MAX;
   }

   public static void dequeue() {
      if(myqueue.size() > 0) {
        int front = (int)myqueue.get(0);
        myqueue.remove(0);
        System.out.println("removed item : "+front);
      } else {
        System.out.println("* myqueue is empty *");
      }
   }

   public static void push(int item) {
      if(isFull() == false) {
        myqueue.add(item);
      } else {
```

```
            System.out.println("* myqueue is full *");
        }
    }

    public static void main(String[] args)
    {
        int[] arr1 = new int[]{10,20,200,50,-123};

        System.out.println("=> pushing values onto
        myqueue:");
        for(int num : arr1) {
          push(num);
          System.out.println("myqueue: "+myqueue);
        }

        System.out.println("=> dequeue values from myqueue: ");
        while(myqueue.size() > 0) {
          dequeue();
          System.out.println("myqueue: "+myqueue);
        }
    }
}
```

Listing 7.2 starts by initiating my queue as an empty list and assigning the value 4 to the variable `MAX`, which is the maximum number of elements that the queue can contain (obviously you can change this value).

The next portion of Listing 7.2 defines several methods: the `isEmpty` function that returns `True` if the length of `myqueue` is 0 (and `False` otherwise), followed by the function `isFull` that returns `True` if the length of `myqueue` is `MAX` (and `False` otherwise).

The next portion of Listing 7.2 defines the method `dequeue()` that invokes the `pop()` method in order to remove the front element of `myqueue`, provided that `myqueue` is not empty. Next, the method `push()` invokes the `append()` method in order to add a new element to the back of `myqueue`, provided that `myqueue` is not full.

The final portion of Listing 7.2 invokes the push() method to append various numbers to `myqueue`, followed by multiple invocations of the `dequeue()` method to remove elements from the front of the queue. Launch the code in Listing 7.2 and you will see the same output at Listing 7.1.

CREATING A QUEUE USING AN ARRAY

Listing 7.3 displays the contents of `QueueArray.java` that illustrates how to use a `Java` class in order to define a queue using an array.

LISTING 7.3: QueueArray.java

```
import java.util.ArrayList;

public class QueueArray
{
   static int MAX = 4;
   static ArrayList myqueue = new ArrayList();

   public static Boolean isEmpty() {
     return myqueue.size() == 0;
   }

   public static Boolean isFull() {
     return myqueue.size() == MAX;
   }

   public static void dequeue() {
      if(myqueue.size() > 0) {
        int front = (int)myqueue.get(0);
        myqueue.remove(0);
        System.out.println("removed item : "+front);
      } else {
        System.out.println("* myqueue is empty *");
      }
   }

   public static void push(int item) {
      if(isFull() == false) {
        myqueue.add(item);
      } else {
        System.out.println("* myqueue is full *");
      }
   }

   public static void main(String[] args)
   {
      int[] arr1 = new int[]{10,20,200,50,-123};

      System.out.println("=> pushing values onto myqueue:");
      for(int num : arr1) {
        push(num);
        System.out.println("myqueue: "+myqueue);
      }

      System.out.println("=> dequeue values from myqueue: ");
      while(myqueue.size() > 0) {
        dequeue();
        System.out.println("myqueue: "+myqueue);
      }
   }
}
```

Listing 7.3 starts by initializing the variables MAX (for the maximum size of the queue), myqueue (which is an array-based queue), along with the integers lpos and rpos that are the index positions of the first element and the last element, respectively, of the queue.

The next portion of Listing 7.3 defines the familiar methods isEmpty() and isFull() that you have seen in previous code samples. However, the dequeue() function has been modified to handle cases in which elements are popped from myqueue: each time this happens, the variable lpos is incremented by 1. Note that this code block is executed only when lpos is less than rpos: otherwise, the queue is empty.

The final portion of Listing 7.3 initializes the array arr1 with a set of integers, followed by a loop that iterates through the elements of arr1 and invokes the push() function in order to append those elements to myqueue. When this loop finishes execution, another loop invokes the dequeue() function to remove elements from the front of the queue.

Change the value of MAX so that its value is less than, equal to, or greater than the number of elements in the array arr1. Doing so will exhibit different execution paths in the code. Note that numerous print() statements are included in Listing 7.3 that generate verbose output, thereby enabling you to see the sequence in which the code is executed (you can "comment out" those statements later). Launch the code in Listing 7.3 and you will see the following output:

```
=> pushing values onto myqueue:
myqueue: [10]
myqueue: [10, 20]
myqueue: [10, 20, 200]
myqueue: [10, 20, 200, 50]
* myqueue is full *
myqueue: [10, 20, 200, 50]
=> dequeue values from myqueue:
removed item : 10
myqueue: [20, 200, 50]
removed item : 20
myqueue: [200, 50]
removed item : 200
myqueue: [50]
removed item : 50
myqueue: []
```

OTHER TYPES OF QUEUES

In addition to the queues that you have seen thus far in this chapter, there are several other types of queues, as listed below:

- circular queues
- priority queues
- dequeues

A *circular queue* is a queue whose "head" is the same as its "tail." A priority queue is a queue in which elements are assigned a numeric priority. A dequeue is a queue in which elements can be added as well as removed from both ends of the queue. Search online for code samples that implement these types of queues.

This concludes the portion of the chapter pertaining to queues. The remainder of this chapter discusses the stack data structure, which is based on a LIFO (last-in-first-out) structure instead of a FIFO structure of a queue.

WHAT IS A STACK?

In general terms, a stack consists of a collection of objects following the LIFO principle. By contrast, a stack follows the LIFO principle.

As a simple example, consider an elevator that has one entrance: the last person who enters the elevator is the first person who exits the elevator. Thus, the order in which people exit an elevator is the reverse of the order in which people enter an elevator.

Another analogy that might help you understand the concept of a stack is the stack of plates in a cafeteria:

1. A plate can be added to the top of the stack if the stack is not full.
2. A plate can be removed from the stack if the stack is not empty.

Based on the preceding observations, a stack has a maximum size `MAX` and a minimum size of 0.

Use Cases for Stacks

The following list contains use applications and use cases for stack-based data structures:

- recursion
- keeping track of function calls
- evaluation of expressions
- reversing characters
- servicing hardware interrupts
- solving combinatorial problems using backtracking

Operations With Stacks

Earlier in this chapter you saw `Java` methods to perform operations on queues; in an analogous fashion, we can define a stack in terms of the following methods:

- `isEmpty()` returns True if the stack is empty
- `isFull()` returns True if the stack is full

- `stackSize()` returns the number of elements in the stack
- `push(item)` adds an element to the "top" of the stack if the stack is not full
- `pop()` removes the top-most element of the stack if the stack is not empty

In order to ensure that there is no overflow (too big) or underflow (too small), we must always invoke `isEmpty()` before popping an item from the stack and always invoke `isFull()` before "pushing" an item onto the stack. The same methods (with different implementation details) are relevant when working with queues.

WORKING WITH STACKS

Listing 7.4 displays the contents of `MyStack.java` that illustrates how to use the Java ArrayList class to define a stack and perform various operations on the stack.

LISTING 7.4: MyStack.java

```
import java.util.ArrayList;

public class MyStack
{
   static int MAX = 4; // 100
   static ArrayList mystack = new ArrayList();
   static int[] arr1 = new int[]{10,20,-123,200,50};

   public static Boolean isEmpty()
   {
      return mystack.size() == 0;
   }

   public static Boolean isFull()
   {
      return mystack.size() == MAX;
   }

   public static void push(int item)
   {
      if(isFull() == false) {
        mystack.add(item);
        System.out.println("appended stack item: "+item);
      } else {
         System.out.println("stack is full: cannot append
         "+item);
         System.out.println("Current stack size: "+mystack.
         size());
      }
   }
```

```
    public static void main(String[] args)
    {
       System.out.println("MAX: "+MAX);
       System.out.println("empty list: "+isEmpty());
       System.out.println("full  list: "+isFull());
       System.out.println();

       System.out.println("=> pushing values onto mystack:");
       for(int item : arr1) {
          push(item);
       }
       System.out.println();

       System.out.println("=> popping values from mystack:");
       while(mystack.size() > 0) {
          System.out.println("Current stack size: "+mystack.
          size());
          if(isEmpty() == false) {
            int item = (int)mystack.remove(0);
            System.out.println("removed item: "+item);
          } else {
            System.out.println("stack is empty: cannot pop");
          }
       }

    }
}
```

Listing 7.4 is very similar to Listing 7.1, except that we are working with a stack instead of a queue. In particular, Listing 7.4 starts by initializing `mystack` as an empty list and assigning the value 3 to the variable `MAX`, which is the maximum number of elements that the stack can contain (obviously you can change this number).

The next portion of Listing 7.4 defines several methods: the `isEmpty` method that returns `True` if the length of `mystack` is 0 (and `False` otherwise), followed by the method `isFull` that returns `True` if the length of `mystack` is `MAX` (and false otherwise).

The next portion of Listing 7.4 defines the method `dequeue` that invokes the `pop()` method in order to remove the front element of `mystack`, provided that `mystack` is not empty. Next, the method `push()` invokes the `append()` method in order to add a new element to the top of `mystack`, provided that `myqueue` is not full.

The final portion of Listing 7.4 invokes the `push()` method to append various numbers to `mystack`, followed by multiple invocations of the dequeue() method to remove elements from the top of `mystack`. Launch the code in Listing 7.4 and you will see the following output:

```
pushing values onto mystack:
mystack: [10]
mystack: [10, 20]
mystack: [10, 20, -123]
* mystack is full *
mystack: [10, 20, -123]
* mystack is full *
mystack: [10, 20, -123]

popping values from mystack:
mystack: [10, 20]
mystack: [10]
mystack: []
* mystack is empty *
mystack: []
* mystack is empty *
mystack: []
```

Listing 7.5 displays the contents of `MyStack2.java` that illustrates how to define a stack and perform various operations on the stack.

LISTING 7.5: MyStack2.java

```
import java.util.ArrayList;

public class MyStack2
{
   static int MAX = 3; // 100
   static ArrayList mystack = new ArrayList();
   static int[] arr1 = new int[]{10,20,-123,200,50};

   public static Boolean isEmpty()
   {
      return mystack.size() == 0;
   }

   public static Boolean isFull()
   {
      return mystack.size() == MAX;
   }

   public static void push(int item)
   {
      if(isFull() == false) {
        mystack.add(item);
      } else {
         System.out.println("stack is full: cannot push");
      }
   }

   public static void main(String[] args)
   {
      System.out.println("MAX: "+MAX);
```

```
        System.out.println("empty list: "+isEmpty());
        System.out.println("full  list: "+isFull());

        System.out.println("=> pushing values onto mystack:");
        for(int item : arr1) {
           System.out.println("push item: "+item);
        }

        System.out.println("=> popping values from mystack:");
        for(int item : arr1) {
           //System.out.println("popping item: "+item);
        }
    }
}
```

Listing 7.5 is straightforward because it's a direct counterpart to Listing 7.2: the latter involves a queue whereas the former involves a stack. Launch the code in Listing 7.5 and you will see the following output:

```
pushing values onto mystack:
mystack: [0]
mystack: [0, 1]
mystack: [0, 1, 2]
* mystack is full *
mystack: [0, 1, 2]
* mystack is full *
mystack: [0, 1, 2]

popping values from mystack:
mystack: [0, 1]
mystack: [0]
mystack: []
* mystack is empty *
mystack: []
* mystack is empty *
mystack: []
```

TASK: REVERSE AND PRINT STACK VALUES

Listing 7.6 displays the contents of `ReverseStack.java` that illustrates how to define a stack and print its contents in reverse order.

LISTING 7.6: ReverseStack.java

```
import java.util.ArrayList;

public class ReverseStack
{
   static int MAX = 4; // 100
   static ArrayList mystack = new ArrayList();
   static int[] arr1 = new int[]{10,20,-123,200,50};
```

```
    public static Boolean isEmpty()
    {
       return mystack.size() == 0;
    }

    public static Boolean isFull()
    {
       return mystack.size() == MAX;
    }

    public static void push(int item)
    {
       if(isFull() == false) {
         mystack.add(item);
         System.out.println("appended stack item: "+item);
       } else {
         System.out.println("stack is full: cannot append "+item);
         System.out.println("Current stack size: "+mystack.size());
       }
    }

    public static void main(String[] args)
    {
       System.out.println("MAX: "+MAX);
       System.out.println("empty list: "+isEmpty());
       System.out.println("full  list: "+isFull());
       System.out.println();

       System.out.println("=> pushing values onto mystack:");
       for(int item : arr1) {
          push(item);
       }
       System.out.println();

       ArrayList reversed = new ArrayList();
       System.out.println("=> reversing contents of mystack:");
       while(mystack.size() > 0) {
          System.out.println("Current stack size: "+mystack.
          size());
          if(isEmpty() == false) {
            int item = (int)mystack.remove(0);
            System.out.println("removed item: "+item);
            reversed.add(item);
          } else {
            System.out.println("stack is empty: cannot pop");
          }
       }
       System.out.println("reversed: "+reversed);
    }
}
```

Listing 7.6 contains the code in Listing 7.5, along with a loop that invokes the `push()` method to insert the elements of the `NumPy` array `arr1` (which contains integers) in the variable `mystack`.

After the preceding loop finishes execution, another loop iterates through the elements of `mystack` by invoking the `pop()` method, and in turn appends each element to the array reversed. As a result, the elements in the array reversed are the reverse order of the elements in `mystack`. Launch the code in Listing 7.6 and you will see the following output:

```
MAX: 4
empty list: true
full  list: false

=> pushing values onto mystack:
appended stack item: 10
appended stack item: 20
appended stack item: -123
appended stack item: 200
stack is full: cannot append 50
Current stack size: 4

=> reversing contents of mystack:
Current stack size: 4
removed item: 10
Current stack size: 3
removed item: 20
Current stack size: 2
removed item: -123
Current stack size: 1
removed item: 200
reversed: [10, 20, -123, 200]
```

TASK: DISPLAY THE MIN AND MAX STACK VALUES (1)

Listing 7.7 displays the contents of `StackMinMax.jav` that illustrates how to define a stack and perform various operations on the stack.

LISTING 7.7: StackMinMax.java

```
import java.util.ArrayList;

public class MinMaxStack
{
   static int MAX = 8; // 100
   static int min_val = 99999;
   static int max_val = -99999;
   static ArrayList mystack = new ArrayList();
   static int[] arr1 = new int[]{1000,2000,8000,5000,-1000};
```

```
   public static Boolean isEmpty()
   {
      return mystack.size() == 0;
   }

   public static Boolean isFull()
   {
      return mystack.size() == MAX;
   }

   public static void push(int item)
   {
      if(isFull() == false) {
        mystack.add(item);
        updateMinMaxValues(item);
      } else {
         System.out.println("stack is full: cannot push");
      }
   }

   public static void updateMinMaxValues(int item)
   {
      if(min_val > item)
         min_val = item;

      if(max_val < item)
         max_val = item;
   }

   public static void main(String[] args)
   {
      System.out.println("MAX: "+MAX);
      System.out.println("empty list: "+isEmpty());
      System.out.println("full  list: "+isFull());

      System.out.println("=> pushing values onto mystack:");
      for(int item : arr1) {
         push(item);
         //System.out.println("push item: "+item);
      }

      System.out.println();
      System.out.println("Maximum value: "+max_val);
      System.out.println("Minimum value: "+min_val);
   }
}
```

Listing 7.7 contains the familiar methods `isEmpty()`, `isFull()`, and `pop()` that have been discussed in previous code samples. Notice that the method `pop()` invokes the method `updateMinMaxValues()` each time that an element is removed from the stack. The latter method updates the variables `min_val` and `max_val` to keep track of the smallest and largest elements, respectively, in the stack. Launch the code in Listing 7.7 and you will see the following output:

```
=> Pushing list of values onto mystack:
[ 1000  2000  8000  5000 -1000]

min value: -1000
max value: 8000
```

TASK: REVERSE A STRING USING A STACK

Listing 7.8 displays the contents of `ReverseString.java` that illustrates how to use a stack in order to reverse a string.

LISTING 7.8: ReverseString.java

```
import java.util.ArrayList;

public class ReverseString
{
   static int MAX = 40; // 100
   static ArrayList mystack = new ArrayList();

   public static Boolean isEmpty()
   {
      return mystack.size() == 0;
   }

   public static Boolean isFull()
   {
      return mystack.size() == MAX;
   }

   public static void push(char item)
   {
      if(isFull() == false) {
        mystack.add(item);
        //System.out.println("appended stack item: "+item);
      } else {
        System.out.println("stack is full: cannot append "+item);
      }
   }

   public static void main(String[] args)
   {
      //System.out.println("MAX: "+MAX);
      //System.out.println("empty list: "+isEmpty());
      //System.out.println("full  list: "+isFull());
```

```
        //System.out.println();
        String myString = "abcdxyz";

        System.out.println("original: "+myString);
        //System.out.println("=> pushing characters onto
        mystack:");
        for(int i=0; i<myString.length(); i++) {
           push(myString.charAt(i));
           //System.out.println("current char: "+myString.
           charAt(i));
        }

        // => Append characters to a string:
        //System.out.println("=> reversing contents of mystack:");
        String reversed = "";
        while(mystack.size() > 0) {
           if(isEmpty() == false) {
             //char item = (char)mystack.remove(0);
             char item = (char)mystack.remove(mystack.size()-1);
             //System.out.println("popped last item: "+item);
             reversed = reversed + item;
           } else {
             break;
             //System.out.println("stack is empty: cannot pop");
           }
        }
        System.out.println("reversed: "+reversed);
    }
}
```

Listing 7.8 starts by initializing `mystack` as an empty list, followed by the usual functions `isEmpty()`, `isFull()`, and `pop()` that perform their respective operations. The next portion of Listing 7.8 initializes the variable `my_str` as a string of characters, and then pushes each character onto `mystack`. Next, a for loop removes each element from `mystack` and then appends each element to the string reversed (which initialized as an empty string). Launch the code in Listing 7.8 and you will see the following output:

```
string:   abcdxyz
reversed: zyxdcba
```

TASK: FIND STACK PALINDROMES

Listing 7.9 displays the contents of `StackPalindrome.java` that illustrates how to define a stack and perform various operations on the stack.

LISTING 7.9: StackPalindrome.java

```
import java.util.ArrayList;

public class StackPalindrome
```

```
{
   static int MAX = 40; // 100
   static ArrayList mystack = new ArrayList();

   public static Boolean isEmpty()
   {
      return mystack.size() == 0;
   }

   public static Boolean isFull()
   {
      return mystack.size() == MAX;
   }

   public static void push(char item)
   {
      if(isFull() == false) {
        mystack.add(item);
        //System.out.println("appended stack item: "+item);
      } else {
        System.out.println("stack is full: cannot append "+item);
        //System.out.println("Current stack size:
        "+mystack.size());
      }
   }

   public static void main(String[] args)
   {
      Boolean palindrome = true;
      String[] myStrings = new String[]
         {"abc", "bob", "radar", "abcdefdcba"};
      // compare characters from opposite ends of the ArrayList:
      for(String currStr : myStrings)
      {
         palindrome = true;
         char item1, item2;
         for(int idx=0; idx<currStr.length()/2; idx++)
         {
            item1 = currStr.charAt(idx);
            item2 = currStr.charAt(currStr.length()-idx-1);
            if(item1 != item2) {
               palindrome = false;
               break;
            }
         }
         System.out.println("string:     "+currStr);
         System.out.println("palindrome: "+palindrome);
         System.out.println();
      }
   }
}
```

Listing 7.9 starts by defining the variable mystic as an instance of the ArrayList class, which is a structure that represents a stack. In addition, Listing 7.9 contains the methods isEmpty(), isFull(), push(), and main() that check for an empty stack, check for a full stack, place a new item on the stack, and the main() method, respectively.

The isEmpty() method returns a Boolean value based on whether or not the stack has size 0, and isFull() method returns a Boolean value based on whether or not the stack has size MAX, which is the maximum allowable size for the stack. The push() method first checks of the isFull() method returns false; if so, an item is added to the stack. Otherwise, a message is printed that indicates the stack is full.

The main() method initializes the variable myStrings as an array of strings, followed by a loop that iterates through the elements of myStrings. During each iteration, elements from opposite ends of a given string are compared: if they differ, then the string is not a palindrome, the variable palindrome is set to false and an early exit from the loop is performed via a break statement. After the loop has finished execution, the current string is displayed along with a message indicating whether or not the string is a palindrome. Launch the code in Listing 7.9 and you will see the following output:

```
string:     abc
palindrome: false

string:     bob
palindrome: true

string:     radar
palindrome: true

string:     abcdefdcba
palindrome: false
```

TASK: BALANCED PARENTHESES

Listing 7.10 displays the contents of the file `ArrayBalancedParens.java` that illustrates how to use a `NumPy` array in order to determine whether or not a string contains balanced parentheses.

LISTING 7.10: ArrayBalancedParens.java

```
import java.util.ArrayList;

public class StackBalancedParentheses
{
   static int MAX = 100;
   static ArrayList mystack = new ArrayList();

   public static Boolean isEmpty()
   {
```

```
    return mystack.size() == 0;
  }

  public static Boolean isFull()
  {
    return mystack.size() == MAX;
  }

  public static void push(char item)
  {
    if(isFull() == false) {
      mystack.add(item);
    } else {
      System.out.println("stack is full: cannot append "+item);
    }
  }

  public static Boolean pairInBalanPairs(String[] balan_
  pairs, String str)
  {
    for(String elem : balan_pairs) {
      if(elem.contains(str))
        return true;
    }
    return false;
  }

  public static Boolean checkBalanced(String my_expr)
  {
    String left_chars  = new String("([{");
    String right_chars = new String(")]}");
    String[] balan_pairs = new String[]
    {"'()'","'[]'","'{}'"};
    ArrayList my_stack = new ArrayList();

    System.out.println("checking string: "+my_expr);
    for(int idx=0; idx<my_expr.length(); idx++) {
      char char1      = my_expr.charAt(idx);
      String char1Str = String.valueOf(char1);
      //System.out.println("char: "+char1);

      if(left_chars.contains(char1Str)) {
        my_stack.add(char1Str);
        //System.out.println("appended to my_stack: "+my_stack);
      } else if (right_chars.contains(char1Str)) {
        if(my_stack.size() > 0) {
          String top1Str = (String)my_stack.get(my_stack.
          size()-1);
          String two_chars = top1Str + char1Str;
          //System.out.println("=> two_chars: "+two_chars);
```

```
          if(pair_in_balan_pairs(balan_pairs, two_chars)
          == true) {
            my_stack.remove(my_stack.size()-1);
            if(my_stack.size() == 0)
              my_stack = new ArrayList();
              //System.out.println("new stack: "+my_stack);
            continue;
          } else {
            //System.out.println("non-match: "+char1);
            break;
          }
        } else {
          System.out.println("empty stack: invalid string
          "+my_expr);
          return false;
        }
      } else {
        System.out.println("invalid character: "+char1);
        return false;
      }
    }

    return (my_stack.size() == 0);
  }

  public static void main(String[] args)
  {
     String[] exprs = new String[]{"[()]{}{[()()]()}",
                                   "[()]{}{[()()]()]",
                                   "(()]{}{[()()]()]",
                                   "(())(()()()()()]"};

     for(String expr : exprs) {
       if(checkBalanced(expr) == true)
         System.out.println("=> balanced:     "+expr);
       else
         System.out.println("=> unbalanced:   "+expr);
       System.out.println();
     }
  }
}
```

Listing 7.10 is the longest code sample in this chapter that also reveals the usefulness of combining a stack with recursion in order to solve the task at hand, which is to determine which strings comprise balanced parentheses.

Listing 7.10 starts with the method `checkBalanced()` that takes a string called `my_expr` as its lone parameter. Notice the way that the following variables are initialized:

```
left_chars = "([{"
right_chars = ")]}"
balan_pairs = np.array(["()","[]", "{}"])
```

The variable `left_chars` and `right_chars` contain the left-side parentheses and right-side parentheses that are permissible in a well-balanced string. Next, the variable `balan_pairs` is an array of three strings that represent a balanced pair of round parentheses, square parentheses, and curly parentheses, respectively.

The key idea for this code sample involves two actions and a logical comparison, as listed here:

1. Whenever a *left* parenthesis is encountered in the current string, this parenthesis is pushed onto a stack.
2. Whenever a *right* parenthesis is encountered in the current string, we check the top of the stack to see if it equals the corresponding left parenthesis.
3. If the comparison in #2 is true, we pop the top element of the stack.
4. If the comparison in #2 is false, the string is unbalanced.

Repeat the preceding sequence of steps until we reach the end of the string: if the stack is empty, the expression is balanced, otherwise the expression is unbalanced.

For example, suppose we the string `my_expr` is initialized as "()". The first character is "(", which is a left parenthesis: step #1 above tells us to push "(" onto our (initially empty) stack called `mystack`. The next character in `my_expr` is ")", which is a right parenthesis: step #2 in the previous list tells us to compare ")" with the element in the top of the stack, which is "(". Since "(" and ")" constitute a balanced pair of parentheses, we pop the element "(" from the stack. We have also reached the end of `my_expr`, and since the stack is empty, we conclude that "()" is a balanced expression (which we knew already).

Now suppose that we the string `my_expr` is initialized as "((". The first character is "(", which is a left parenthesis, and step #1 in the previous list tells us to push "(" onto our (initially empty) stack called `mystack`. The next character in `my_expr` is "(", which is a left parenthesis: step #1 in the previous list tells us to push "(" onto the stack. We have reached the end of `my_expr` and since the stack is non-empty, we have determined that `my_expr` is unbalanced.

As a third example, suppose that we the string `my_expr` is initialized as "(()". The first character is "(", which is a left parenthesis: step #1 in the previous list tells us to push "(" onto our (initially empty) stack called `mystack`. The next character in `my_expr` is "(", which is a left parenthesis: step #1 in the previous list tells us push another "(" onto the stack. The next character in `my_expr` is ")", and step #2 tells us to compare ")" with the element in the top of the stack, which is "(". Since "(" and ")" constitute a balanced pair of parentheses, we pop the top-most element of the stack. At this point we have reached the end of `my_expr`, and since the stack is non-empty, we know that `my_expr` is unbalanced.

Try tracing through the code with additional strings consisting of a sequence of parentheses. After doing so, the code details in Listing 7.13 will become much simpler to understand.

The next code block in Listing 7.10 initializes the variable `my_expr` as an array of strings, each of which consists of various parentheses (round, square, and curly). The next portion is a loop that iterates through the elements of `my_expr` and in turn invokes the function `checkBalanced()` to determine which ones (if any) comprise balanced parentheses. Launch the code in Listing 7.10 and you will see the following output:

```
checking string: [()]{}{[()()]()}
=> balanced:     [()]{}{[()()]()}

checking string: [()]{}{[()()]()]
=> unbalanced:   [()]{}{[()()]()]

checking string: (()]{}{[()()]()]
=> unbalanced:   (()]{}{[()()]()]

checking string: (())(()()()()()]
=> unbalanced:   (())(()()()()()]
```

Consider enhancing Listing 7.13 so that the invalid character is displayed for strings that consists of unbalanced parentheses.

TASK: TOKENIZE ARITHMETIC EXPRESSIONS

The code sample in this section is a prelude to the task in the next section that involves parsing arithmetic expressions: in fact, the code in Listing 7.11 is included in the code in Listing 7.11. The rationale for the inclusion of a separate code sample is to enable you to tokenize expressions that might not be arithmetic expressions.

You need to download the zip file `apache-commons-lang.jar.zip`, uncompress the contents of the zip file, and update the environment variable `CLASSPATH` as follows:

- *http://www.java2s.com/Code/Jar/a/Downloadapachecommonslangjar.htm*
- `export CLASSPATH=$CLASSPATH:apache-commons-lang.jar`

Listing 7.11 displays the contents of `TokenizeExpr.java` that illustrates how to tokenize an arithmetic expression and also remove whitespaces and tab characters.

LISTING 7.11: TokenizeExpr.java

```
import org.apache.commons.lang.text.StrTokenizer;
```

```
public class TokenizeExpr
{
   public static void main(String[] args)
   {
      String[] exprs = new String[]{"         2789  *  3+7-8-  9",
                                   "4 /2  +  1",
                                   "2  -  3  +  4      "};

      for(String expr : exprs) {
        String parsed = "";
        System.out.println("String: "+expr);
        StrTokenizer tokenizer = new StrTokenizer(expr, " ");
        while (tokenizer.hasNext()) {
           String token = tokenizer.nextToken().trim();
           //System.out.println("token: "+token);
           parsed = parsed + " " + token;
        }
        System.out.println("Parsed: "+parsed);
        System.out.println();
      }
   }
}
```

Listing 7.11 contains a main() method that initializes the variable exprs as an array of strings, each of which is an arithmetic expression. Next, a loop iterates through the elements of exprs. During each iteration, the variable tokenizer is instantiated as an instance of the Java StrTokenizer class, after which the hasNext() method of the variable tokenizer is invoked in order to iterate through the tokens of the current string in the exprs variable. During each iteration, the current token is prepended to the string variable parse and then its contents are displayed. Launch the code in Listing 7.11 and you will see the following output that has removed the extra whitespaces from each string in the exprs variable:

```
String:          2789  *  3+7-8-  9
Parsed:  2789 * 3+7-8- 9

String: 4 /2  +  1
Parsed:  4 /2 + 1

String: 2  -  3  +  4
Parsed:  2 - 3 + 4
```

TASK: CLASSIFY TOKENS IN ARITHMETIC EXPRESSIONS

Evaluating arithmetic expressions is an interesting task because there are various ways to approach this problem. First, some people would argue that the "real" way to solve this task involves levers and parsers. However, the purpose of this task is to familiarize you with parsing strings, after which you will be better equipped to do so with nonarithmetic expressions.

Listing 7.12 displays the contents of `ClassifyTokens.java` that illustrates how to parse and evaluate an arithmetic expression using a stack.

LISTING 7.12: ClassifyTokens.java

```
import java.util.ArrayList;
import org.apache.commons.lang.text.StrTokenizer;

public class ClassifyTokens
{
   //parse a string into tokens and classify them
   public static void tokenize(String my_expr)
   {
     ArrayList tokens_types = new ArrayList();
     String[] math_symbols = new String[] {"*","/","+","-"};
     String all_digits = new String("0123456789");
     String oper = "";
     String int_str1 = "";

     my_expr = parseString(my_expr);
     System.out.println("STRING: "+my_expr);

     int idx=0;
     while(idx<my_expr.length()) {
       // 1) extract integer:
       int_str1 = "";
       char token = my_expr.charAt(idx);
       String tokenStr = String.valueOf(token);

       System.out.println("extract number:");
       while(idx<my_expr.length())
       {
          System.out.println("tokenStr: "+tokenStr+ " idx:
          "+idx);
          if(all_digits.contains(tokenStr)) {
             int_str1 += tokenStr;
             idx += 1;
          } else {
            break;
          }

          if( idx == my_expr.length()) {
             System.out.println("Reached end of string");
             break;
          }

          token = my_expr.charAt(idx);
          tokenStr = String.valueOf(token);
       }
       tokens_types.add("token: "+int_str1 + " type: NUMERIC");
       System.out.println("=> int_str1: "+int_str1);
```

```
    System.out.println();

    // 2) check for an arithmetic symbol: *, /, +, or -
    if(substr_in_array_strings(math_symbols, tokenStr)
    == true) {
      oper = tokenStr;
      System.out.println("=> found operator: "+oper);
      System.out.println();
      tokens_types.add("token: "+oper + " type: SYMBOL");
    }
    idx += 1;
  }
  System.out.println("SUMMARY:");
  for(int i=0; i<tokens_types.size(); i++)
  {
    System.out.println("string/type: "+tokens_types.get(i));
  }
  System.out.println("-------------------\n");
}

public static Boolean substrInArrayStrings(String[]
my_array,String str)
{
  for(String elem : my_array) {
    if(elem.contains(str))
      return true;
  }
  return false;
}

public static String parseString(String expr)
{
   String parsed = "";
   //System.out.println("String: "+expr);
   StrTokenizer tokenizer = new StrTokenizer(expr, " ");
   while (tokenizer.hasNext()) {
      String token = tokenizer.nextToken().trim();
      //System.out.println("token: "+token);
      parsed = parsed + token;
   }
   System.out.println();
   return parsed;
}

public static void main(String[] args)
{
   String[] exprs = new String[]{"          2789   *
   3+7-8-  9",
                                          "4 /2  +  3",
                                          "4 /2  +  1",
                                          "9  -  3  +  4     "};
```

```
        for(String expr : exprs) {
           tokenize(expr);
        }
     }
}
```

Listing 7.12 starts by defining the variable tokens_types as an instance of the Java ArrayList class, followed by the variable math_symbols that contains a 7.12 also defines the methods `tokenize()`, `substrInArrayString()`, and parseString() that tokenizes a string, checks a string for a substring, and parses a string, respectively.

The code for parseString() is the same as the method in the TokenizeExpr class in the previous section. The `substrInArrayString()` method checks if a given string is an element of an array of strings. The parseString() method is also the same as the method in the TokenizeExpr class in the previous section.

The main() method that initializes the variable exprs as an array of strings, each of which is an arithmetic expression. Next, a loop iterates through the elements of exprs. During each iteration, the method tokenize() is invoked with the current element in the exprs array. Now Launch the code in Listing 7.9 and you will see the following output:

```
STRING: 2789*3+7-8-9
extract number:
tokenStr: 2 idx: 0
tokenStr: 7 idx: 1
tokenStr: 8 idx: 2
tokenStr: 9 idx: 3
tokenStr: * idx: 4
=> int_str1: 2789

=> found operator: *

extract number:
tokenStr: 3 idx: 5
tokenStr: + idx: 6
=> int_str1: 3

=> found operator: +

extract number:
tokenStr: 7 idx: 7
tokenStr: - idx: 8
=> int_str1: 7

=> found operator: -

extract number:
tokenStr: 8 idx: 9
tokenStr: - idx: 10
```

```
=> int_str1: 8

=> found operator: -

extract number:
tokenStr: 9 idx: 11
Reached end of string
=> int_str1: 9

SUMMARY:
string/type: token: 2789 type: NUMERIC
string/type: token: * type: SYMBOL
string/type: token: 3 type: NUMERIC
string/type: token: + type: SYMBOL
string/type: token: 7 type: NUMERIC
string/type: token: - type: SYMBOL
string/type: token: 8 type: NUMERIC
string/type: token: - type: SYMBOL
string/type: token: 9 type: NUMERIC
-------------------


STRING: 4/2+3
extract number:
tokenStr: 4 idx: 0
tokenStr: / idx: 1
=> int_str1: 4

=> found operator: /

extract number:
tokenStr: 2 idx: 2
tokenStr: + idx: 3
=> int_str1: 2

=> found operator: +

extract number:
tokenStr: 3 idx: 4
Reached end of string
=> int_str1: 3

SUMMARY:
string/type: token: 4 type: NUMERIC
string/type: token: / type: SYMBOL
string/type: token: 2 type: NUMERIC
string/type: token: + type: SYMBOL
string/type: token: 3 type: NUMERIC
-------------------

STRING: 4/2+1
```

```
extract number:
tokenStr: 4 idx: 0
tokenStr: / idx: 1
=> int_str1: 4

=> found operator: /

extract number:
tokenStr: 2 idx: 2
tokenStr: + idx: 3
=> int_str1: 2

=> found operator: +

extract number:
tokenStr: 1 idx: 4
Reached end of string
=> int_str1: 1

SUMMARY:
string/type: token: 4 type: NUMERIC
string/type: token: / type: SYMBOL
string/type: token: 2 type: NUMERIC
string/type: token: + type: SYMBOL
string/type: token: 1 type: NUMERIC
--------------------


STRING: 9-3+4
extract number:
tokenStr: 9 idx: 0
tokenStr: - idx: 1
=> int_str1: 9

=> found operator: -

extract number:
tokenStr: 3 idx: 2
tokenStr: + idx: 3
=> int_str1: 3

=> found operator: +

extract number:
tokenStr: 4 idx: 4
Reached end of string
=> int_str1: 4

SUMMARY:
string/type: token: 9 type: NUMERIC
string/type: token: - type: SYMBOL
```

```
string/type: token: 3 type: NUMERIC
string/type: token: + type: SYMBOL
string/type: token: 4 type: NUMERIC
-------------------
```

INFIX, PREFIX, AND POSTFIX NOTATIONS

There are three well-known and useful techniques for representing arithmetic expressions: infix notation, prefix notation, and postfix notation.

Infix notation involves specifying operators *between* their operands, which is the typical way that we write arithmetic expressions (example: 3+4*5).

Prefix notation (also called Polish notation) involves specifying operators *before* their operands. For example, the infix expression 3+4 has the sequence + 3 4 as its equivalent prefix expression.

Postfix notation (also called reverse Polish notation) involves specifying operators *after* their operands, examples of which are here:

4*5 becomes 4 5 *
a*b + c*d becomes a b * c d * +
a * b – (c + d) becomes a b * c d + –
3+4*5 becomes 3 4 5 * +

You can use a stack to perform operations on postfix operations that involve pushing numeric (or variable) values onto a stack, and when a binary operator is found in the postfix expression, apply that operator to the top two elements from the stack and replace them with the resulting operation. As an example, let's look at the contents of a stack that calculate the expression 3+4*5 whose postfix expression is 3 4 5 * +:

Step 1: push 3

3

Step 2: push 4

4

3

Step 3: push 5

5

4

3

Step 4: apply * to the two top elements of the stack

20

3

Step 5: apply + to the two top elements of the stack

23

Since we have reached the end of the postfix expression, we see that there is one element in the stack, which is the result of evaluating the postfix expression.

The following table contains additional examples of expressions using infix, prefix, and postfix notation.

```
Infix        Prefix       Postfix
x+y          +xy          xy+
x-y          -xy          xy-
x/y          /xy          xy/
x*y          *xy          xy*
x^y          ^yx          yx^

(x+y)*z      *(x+y)z      (x+y)z*
(x+y)*z      *(+xy)z      (xy+)z*
```

Let's look at the following slightly more complex infix expression (note the "/" that is shown in bold):

```
[[x+(y/z)-d]^2]/(x+y)
```

We will perform an iterative sequence of steps to convert this infix expression to a prefix expression by applying the definition of infix notation to the top-level operator. In this example, the top-level operator is the "/" symbol that is shown in bold. We need to place this "/" symbol in the left-most position, as shown here (and notice the "^" symbol shown in bold):

```
/[[x+y/z-d]^2](x+y)
```

Now we need to place this "^" symbol immediately to the left of the second left square bracket, as shown here (and notice the "/" shown in bold):

```
/[^[x+(y/z)-d]2](+xy)
```

Now we need to place this "/" symbol immediately to the left of the variable y, as shown here (and notice the "+" shown in bold):

```
/[^[x+(/yz)-d]2](+xy)
```

Now we need to place this "+" symbol immediately to the left of the variable x, as shown here (and notice the "/" shown in bold):

```
/[^[+x(/yz)-d]2](+xy)
```

Now we need to place this "/" symbol immediately to the left of the variable x, as shown here, which is now an infix expression:

```
/[^[-(+(/yz))d]2](+xy)
```

Perform a search offline for more examples of prefix and postfix expressions as well as code samples that implement those expressions using a Java data structure.

SUMMARY

This chapter started with an introduction to queues, along with real-world examples of queues. Next, you learned about several functions that are associated with a queue, such as `isEmpty()`, `isFull()`, `push()`, and `dequeue()`.

Next you learned about stacks, which are LIFO data structures, along with some `Java code` samples that show you how to perform various operations on stacks. Some examples include reversing the contents of a stack and also determining whether or not the contents of a stack form a palindrome.

In the final portion of this chapter, you learned how to determine whether or not a string consists of well-balanced round parentheses, square brackets, and curly braces; and also how to convert infix notation to postfix notation.

CHAPTER 8

ChatGPT and GPT-4

This chapter contains information about the main features of ChatGPT and GPT-4, as well as some of their competitors.

The first portion of this chapter starts with information generated by ChatGPT regarding the nature of generative AI and conversational AI versus generative AI. According to ChatGPT, it's also true that ChatGPT, GPT-4, and Dall-E are also included in generative AI. This section also discusses ChatGPT and some of its features, as well as alternatives to ChatGPT.

The third portion of this chapter introduces some of the features of GPT-4 that power ChatGPT. You will also learn about some competitors of GPT-4, such as Llama-2 (Meta) and Bard (Google).

WHAT IS GENERATIVE AI?

Generative AI refers to a subset of artificial intelligence models and techniques that are designed to generate new data samples that are similar in nature to a given set of input data. The goal is to produce content or data that wasn't part of the original training set but is coherent, contextually relevant, and in the same style or structure.

Generative AI stands apart in its ability to create and innovate, as opposed to merely analyzing or classifying. The advancements in this field have led to breakthroughs in creative domains and practical applications, making it a cutting-edge area of AI research and development.

Key Features of Generative AI

The following bullet list contains key features of generative AI, followed by a brief description for each bullet item:

- data generation
- synthesis
- learning distributions

Data generation refers to the ability to create new data points that are not part of the training data but that resemble it. This can include text, images, music, videos, or any other form of data.

Synthesis indicates that generative models can blend various inputs to generate outputs that incorporate features from each input, like merging the styles of two images.

Learning distributions means that generative AI models aim to learn the probability distribution of the training data so they can produce new samples from that distribution.

Popular Techniques in Generative AI

Generative adversarial networks (GANs): GANs consist of two networks, a generator and a discriminator, that are trained simultaneously. The generator tries to produce fake data, while the discriminator tries to distinguish between real data and fake data. Over time, the generator gets better at producing realistic data.

Variational autoencoders (VAEs): VAEs are probabilistic models that learn to encode and decode data in a manner such that the encoded representations can be used to generate new data samples.

Recurrent neural networks (RNNs): RNNs are used primarily for sequence generation such as text or music.

What Makes Generative AI Different

Creation versus classification: While most traditional AI models aim to classify input data into predefined categories, generative models aim to create new data.

Unsupervised Learning: Many generative models, especially GANs and VAEs, operate in an unsupervised manner, meaning they don't require labeled data for training.

Diverse outputs: Generative models can produce a wide variety of outputs based on learned distributions, making them ideal for tasks like art generation, style transfer, and more.

Challenges: Generative AI poses unique challenges, such as mode collapse in GANs or ensuring the coherence of generated content.

Furthermore, there are numerous areas that involve generative AI applications, some of which are listed in the following bullet list:

- art and music Creation
- data augmentation
- style transfer
- text generation
- image synthesis

Art and music creation include generating paintings, music, or other forms of art.

Data augmentation involves creating additional data for training models, especially when the original dataset is limited.

Style transfer refers to applying the style of one image to the content of another.

Text generation is a very popular application of generative AI that involves creating coherent and contextually relevant text.

Image synthesis is another popular area of generative AI that involves generating realistic images, faces, or even creating scenes for video games.

Drug discovery is a very important facet of generative AI that pertains to generating molecular structures for new potential drugs.

CONVERSATIONAL AI VERSUS GENERATIVE AI

Both conversational AI and generative AI are prominent subfields within the broader domain of artificial intelligence. However, these subfields have a different focus regarding their primary objective, the technologies that they use, and applications.

More about the differences between conversational AI and generative AI can be found here:

https://medium.com/@social_65128/differences-between-conversational-ai-and-generative-ai-e3adca2a8e9a

The primary differences between the two subfields are in the following list:

- primary objective
- applications
- technologies used
- training and interaction
- evaluation
- data requirements

Primary Objective

The main goal of conversational AI is to facilitate human-like interactions between machines and humans. This includes chatbots, virtual assistants, and other systems that engage in dialogue with users.

The primary objective of generative AI is to create new content or data that wasn't in the training set but is similar in structure and style. This can range from generating images, music, and text to more complex tasks like video synthesis.

Applications

Common applications for conversational AI include customer support chatbots, voice-operated virtual assistants (like Siri or Alexa), and interactive voice response (IVR) systems.

Common applications for generative AI have a broad spectrum of applications such as creating art or music, generating realistic video game environments, synthesizing voices, and producing realistic images or even deep fakes.

Technologies Used

Conversational AI often relies on natural language processing (NLP) techniques to understand and generate human language. This includes intent recognition, entity extraction, and dialogue management.

Generative AI commonly utilizes GANs, variational autoencoders (VAEs), and other generative models to produce new content.

Training and Interaction

While training can be supervised, semi-supervised, or unsupervised, the primary interaction mode for Conversational AI is through back-and-forth dialogue or conversation.

The training process for generative AI, especially with models like GANs, involves iterative processes where the model learns to generate data by trying to fool a discriminator into believing the generated data is real.

Evaluation

Conversational AI evaluation metrics often revolve around understanding and response accuracy, user satisfaction, and the fluency of generated responses.

Generative AI evaluation metrics for models like GANs can be challenging and might involve using a combination of quantitative metrics and human judgment to assess the quality of generated content.

Data Requirements

Data requirements for Conversational AI typically involves dialogue data, with conversations between humans or between humans and bots.

Data requirements for generative AI involve large datasets of the kind of content it is supposed to generate, be it images, text, music, etc.

Although both Conversational AI and generative AI deal with generating outputs, their primary objectives, applications, and methodologies can differ significantly. Conversational AI is geared towards interactive communication with users, while generative AI focuses on producing new, original content.

IS DALL-E PART OF GENERATIVE AI?

DALL-E and similar tools that generate graphics from text are indeed examples of generative AI. In fact, DALL-E is one of the most prominent examples of generative AI in the realm of image synthesis.

Here's a bullet list of generative characteristics of DALL-E, followed by brief descriptions of each bullet item:

- Image Generation
- Learning Distributions
- Innovative Combinations
- Broad Applications
- transformer Architecture

Image Generation is a key feature of DALL-E, which was designed to generate images based on textual descriptions. Given a prompt like "a two-headed flamingo," DALL-E can produce a novel image that matches the description, even if it's never seen such an image in its training data.

Learning Distributions: Like other generative models, DALL-E learns the probability distribution of its training data. When it generates an image, it samples from this learned distribution to produce visuals that are plausible based on its training.

Innovative Combinations: DALL-E can generate images that represent entirely novel or abstract concepts, showcasing its ability to combine and recombine learned elements in innovative ways.

In addition to image synthesis, DALL-E has provided broad application support, in areas like art generation, style blending, and creating images with specific attributes or themes, highlighting its versatility as a generative tool.

DALL-E leverages a variant of the transformer architecture, similar to models like GPT-3, but adapted for image generation tasks.

Other tools that generate graphics, art, or any form of visual content based on input data (whether it's text, another image, or any other form of data) and can produce outputs not explicitly present in their training data are also considered generative AI. They showcase the capability of AI models to not just analyze and classify but to create and innovate.

ARE CHATGPT-3 AND GPT-4 PART OF GENERATIVE AI?

Both ChatGPT-3 and GPT-4 are LLMs that are considered examples of generative AI. They belong to a class of models called "transformers," which are particularly adept at handling sequences of data, such as text-related tasks.

The following bullet list provides various reasons why these LLMs are considered generative, followed by a brief description of each bullet item:

- Text Generation
- Learning Distributions
- Broad Applications
- Unsupervised Learning

Text Generation: These models can produce coherent, contextually relevant, and often highly sophisticated sequences of text based on given prompts. They generate responses that weren't explicitly present in their training data but are constructed based on the patterns and structures they learned during training.

Learning Distributions: GPT-3, GPT-4, and similar models learn the probability distribution of their training data. When generating text, they're essentially sampling from this learned distribution to produce sequences that are likely based on their training.

Broad Applications: Beyond just text-based chat or conversation, these models can be used for a variety of generative tasks like story writing, code generation, poetry, and even creating content in specific styles or mimicking certain authors, showcasing their generative capabilities.

Unsupervised Learning: While they can be fine-tuned with specific datasets, models like GPT-3 are primarily trained in an unsupervised manner on vast amounts of text, learning to generate content without requiring explicit labeled data for every possible response.

In essence ChatGPT-3, GPT-4, and similar models by OpenAI are quintessential examples of generative AI in the realm of NLP and generation.

The next several sections briefly introduce some of the companies that have a strong presence in the AI world.

DEEPMIND

DeepMind has made significant contributions to artificial intelligence, which includes the creation of various AI systems. DeepMind was established in 2010 and became a subsidiary of Google 2014. You can visit the DeepMind home page: *https://deepmind.com/*

DeepMind created the 280GB language model `Gopher` that significantly outperforms its competitors, including `GPT-3`, `J1-Jumbo`, and `MT-NLG`. DeepMind also developed `AlphaFold` that solved a protein folding task in literally 30 minutes that had eluded researchers for ten years. Moreover, DeepMind made `AlphaFold` available for free for everyone in July 2021. DeepMind has made significant contributions in the development of world caliber AI game systems, some of which are discussed in the next section.

DeepMind and Games

DeepMind is the force behind the `AI` systems `StarCraft` and `AlphaGo` that defeated the best human players in `Go` (which is considerably more difficult than chess). These games provide "perfect information," whereas games with "imperfect information" (such as poker) have posed a challenge for machine learning models.

`AlphaGo Zero` (the successor of `AlphaGo`) mastered the game through self-play in less time and with less computing power. `AlphaGo Zero` exhibited extraordinary performance by defeating `AlphaGo` 100–0. Another powerful system is AlphaZero that also used a self-play technique learned to play Go, chess, and shogi, and also achieved SOTA (state of the art) performance results.

By way of comparison, ML models that use tree search are well-suited for games with perfect information. By contrast, games with imperfect information (such as poker) involve hidden information that can be leveraged to devise counter strategies to counteract the strategies of opponents. In particular, AlphaStar is capable of playing against the best players of StarCraft II, and also became the first AI to achieve SOTA results in a game that requires strategic capability.

Player of Games (PoG)

The DeepMind team at Google devised the general-purpose PoG (player of games) algorithm that is based on the following techniques:

- CFR (counterfactual regret minimization)
- CVPN (counterfactual value-and-policy network)
- GT-CFT (growing tree CFR)
- CVPN

The counterfactual value-and-policy network (CVPN) is a neural network that calculates the counterfactuals for each state belief in the game. This is key to evaluating the different variants of the game at any given time.

Growing tree CFR (GT-CFR) is a variation of CFR that is optimized for game-trees trees that grow over time. GT-CFR is based on two fundamental phases, which is discussed in more detail here:

https://medium.com/syncedreview/deepminds-pog-excels-in-perfect-and-imperfect-information-games-advancing-research-on-general-9dbad5c04221

OPENAI

OpenAI is an AI research company that has made significant contributions to AI, including DALL-E and ChatGPT, and its home page is here: *https://openai.com/api/*

OpenAI was founded in San Francisco by Elon Musk and Sam Altman (as well as others), and one of its stated goals is to develop AI that benefits humanity. Given Microsoft's massive investments in and deep alliance with the organization, OpenAI might be viewed as an arm of Microsoft. OpenAI is the creator of the GPT-x series of LLMs (large language models) as well as ChatGPT that was made available on November 30, 2022.

OpenAI made `GPT-3` commercially available via API for use across applications, charging on a per-word basis. `GPT-3` was announced in July 2020 and was available through a beta program. Then in November 2021 OpenAI made `GPT-3` open to everyone, and more details are accessible here:

https://openai.com/blog/api-no-waitlist/

In addition, OpenAI developed `DALL-E` that generates images from text. OpenAI initially did not permit users to upload images that contained realistic faces. Later (Q4/2022) OpenAI changed its policy to allow users to upload faces into its online system. Check the OpenAI Web page for more details. Incidentally, diffusion models have superseded the benchmarks of `DALL-E`.

OpenAI has also released a public beta of `Embeddings`, which is a data format that is suitable for various types of tasks with machine learning, as described here:

https://beta.openai.com/docs/guides/embeddings

OpenAI is the creator of `Codex` that provides a set of models that were trained on NLP. The initial release of Codex was in private beta, and more information is accessible here: *https://beta.openai.com/docs/engines/instruct-series-beta*

OpenAI provides four models that are collectively called their Instruct models, which support the ability of GPT-3 to generate natural language. These models will be deprecated in early January 2024 and replaced with a updated versions of GPT-3, ChatGPT, and GPT-4, as discussed in Chapter 7.

If you want to learn more about the features and services that OpenAI offers, navigate to the following link: *https://platform.openai.com/overview*

COHERE

Cohere is a start-up and a competitor of OpenAI, and its home page is here: *https://cohere.ai/*

Cohere develops cutting-edge `NLP` technology that is commercially available for multiple industries. Cohere is focused on models that perform textual analysis instead of models for text generation (such as GPT-based models). The founding team of Cohere is impressive: CEO Aidan Gomez is one of the co-inventors of the transformer architecture, and CTO Nick Frosst is a protegé of Geoff Hinton.

HUGGING FACE

Hugging Face is a popular community-based repository for open-source `NLP` technology, and its home page is here: *https://github.com/huggingface*

Unlike OpenAI or Cohere, Hugging Face does not build its own `NLP` models. Instead, Hugging Face is a platform that manages a plethora of open-source `NLP` models that customers can fine-tune and then deploy those fine-tuned models. Indeed, Hugging Face has become the eminent location for people to collaborate on `NLP` models, and sometimes described as "GitHub for machine learning and `NLP`."

Hugging Face Libraries

Hugging Face provides three important libraries: datasets, tokenizers, and transformers. The Accelerate library supports `PyTorch` models. The datasets library provides an assortment of libraries for `NLP`. The tokenizers library enables you to convert text data to numeric values.

Perhaps the most impressive library is the transformers library that provides an enormous set of pre-trained `BERT`-based models in order to perform a wide variety of `NLP` tasks. The Github repository is here: *https://github.com/huggingface/transformers*

Hugging Face Model Hub

Hugging Face provides a model hub that provides a plethora of models that are accessible online. Moreover, the website supports online testing of its models, which includes the following tasks:

- masked word completion with BERT
- name Entity Recognition with Electra
- natural language inference with RoBERTa
- question answering with DistilBERT
- summarization with BART
- text generation with GPT-2
- translation with T5

AI21

`AI21` is a company that provides proprietary LLMs via API to support the applications of its customers. The current SOTA model of `AI21` is called `Jurassic-1` (roughly the same size as `GPT-3`), and AI21 also creates its own applications on top of `Jurassic-1` and other models. The current application suite of `AI21` involves tools that can augment reading and writing.

`Primer` is an older competitor in this space, founded two years before the invention of the transformer. The company primarily serves clients in government and defense.

INFLECTIONAI

A more recent company in the `AI` field is `InflectionAI` whose highly impressive founding team includes:

- Reid Hoffman (LinkedIn)
- Mustafa Suleyman (DeepMind cofounder)
- Karen Simonyan (DeepMind researcher)

`InflectionAI` is committed to the challenging task of enabling humans to interact with computers in much the same way that humans communicate with each other.

ANTHROPIC

Anthropic was created in 2021 by former employees of OpenAI. You can access the home page is here: *https://www.anthropic.com/*

Anthropic has significant financial support from an assortment of companies, including Google and Salesforce. As this book goes to print, Anthropic released Claude 2 as a competitor to ChatGPT.

In the meantime, Claude 2 has the ability to summarize as much as 75,000 words of text-based content, whereas ChatGPT currently has a limit of 3,000 words. Moreover, Claude 2 achieved a score of 76.5% on portions of the bar exam and 71% in a coding test. Claude 2 also has a higher rate than `ChatGPT` in terms of providing "clean" responses to queries from users.

This concludes the portion of the chapter regarding the AI companies that are making important contributions in AI. The next section provides a high-level introduction to LLMs.

WHAT IS PROMPT ENGINEERING?

You have already learned about text generators such as GPT-3 and DALL-E 2 from OpenAI, Jurassic from AI21, Midjourney, and Stable Diffusion that can perform text-to-image generation. As such, prompt engineering refers to devising text-based prompts that enable AI-based systems to improve the output that is generated, which means that the output more closely matches whatever users want to produce from AI-systems. By way of analogy, think of prompts as similar to the role of coaches: they offer advice and suggestions to help people perform better in their given tasks.

Since prompts are based on words, the challenge involves learning how different words can affect the generated output. Moreover, it's difficult to predict how systems respond to a given prompt. For instance, if you want to generate a landscape, the difference between a dark landscape and a bright landscape is intuitive. However, if you want a beautiful landscape, how would an AI system generate a corresponding image? As you can surmise, concrete words are easier than abstract or subjective words for AI systems that generate images from text. Just to add more to the previous example, how would you visualize the following:

- a beautiful landscape
- a beautiful song
- a beautiful movie

Although prompt engineering started with text-to-image generation, there are other types of prompt engineering, such as audio-based prompts that interpret emphasized text and emotions that are detected in speech, and sketch-based prompts that generate images from drawings. The most recent focus of attention involves text-based prompts for generating videos, which presents exciting opportunities for artists and designers. An example of image-to-image processing is accessible here:

https://huggingface.co/spaces/fffiloni/stable-diffusion-color-sketch

Prompts and Completions

A *prompt* is a text string that users provide to LLMs, and a *completion* is the text that users receive from LLMs. Prompts assist LLMs in completing a request (task), and they can vary in length. Although prompts can be any text string, including a random string, the quality and structure of prompts affects the quality of completions.

Think of prompts as a mechanism for giving "guidance" to LLMs, or even as a way to "coach" LLMs into providing desired answers. Keep in mind that the number of tokens in a prompt plus the number of tokens in the completion can be at most 2,048 tokens.

Types of Prompts

The following bullet list contains well-known types of prompts for LLMs:

- zero-shot prompts
- one-shot prompts
- few-shot prompts
- instruction prompts

A *zero-shot prompt* contains a description of a task, whereas a *one-shot prompt* consists of a single example for completing a task. As you can probably surmise, *few-shot prompts* consist of multiple examples (typically between 10 and 100). In all cases, a clear description of the task or tasks is recommended: more tasks provide `GPT-3` with more information, which in turn can lead to more accurate completions.

`T0` (for "zero shot") is an interesting LLM: although `T0` is 16 times smaller (11GB) than `GPT-3` (175GB), `T0` has outperformed `GPT-3` on language-related tasks. `T0` can perform well on unseen `NLP` tasks (i.e., tasks that are new to `T0`) because it was trained on a dataset containing multiple tasks.

The following set of links provide the Github repository for `T0`, a Web page for training `T0` directly in a browser, and a 3GB version of `T0`, respectively:

https://github.com/bigscience-workshop/t-zero

As you can probably surmise, `T0++` is based on `T0`, and it was trained with extra tasks beyond the set of tasks on which `T0` was trained.

Another detail to keep in mind: the first three prompts in the preceding bullet list are also called zero-shot learning, one-shot learning, and few-shot learning, respectively.

Instruction Prompts

Instruction prompts are used for fine tuning LLMs, and they specify a format (determined by you) for the manner in which the LLM is expected to conform in its responses. You can prepare your own instruction prompts or you can access prompt template libraries that contain different templates for different tasks, along with different data sets. Various prompt instruction templates are publicly available, such as the following links that provides prompt templates for Llama:

https://github.com/devbrones/llama-prompts

https://pub.towardsai.net/llama-gpt4all-simplified-local-chatgpt-ab7d28d34923

Reverse Prompts

Another technique uses a reverse order: input prompts are answers and the response are the questions associated with the answers (similar to a popular game show). For example, given a French sentence, you might ask the model, "What English text might have resulted in this French translation?"

System Prompts Versus Agent Prompts

The distinction between a system prompt and an agent prompt often comes up in the context of conversational AI systems and chatbot design.

A system prompt is typically an initial message or cue given by the system to guide the user on what they can do or to set expectations about the interaction. It often serves as an introduction or a way to guide users on how to proceed. Here are some examples of system prompts:

```
"Welcome to ChatBotX! You can ask me questions about
weather, news, or sports. How can I assist you today?"

"Hello! For account details, press 1. For technical
support, press 2."

"Greetings! Type 'order' to track your package or 'help'
for assistance."
```

By contrast, an agent prompt is a message generated by the AI model or agent in response to a user's input during the course of an interaction. It's a part of the back-and-forth exchange within the conversation. The agent prompt guides the user to provide more information, clarifies ambiguity, or nudges the user towards a specific action. Here are some examples of agent prompts:

```
User: "I'm looking for shoes."

Agent Prompt: "Great! Are you looking for men's or women's
shoes?"

User: "I can't log in."

Agent Prompt: "I'm sorry to hear that. Can you specify if
you're having trouble with your password or username?"

User: "Tell me a joke."

Agent Prompt: "Why did the chicken join a band? Because it
had the drumsticks!"
```

The fundamental difference between the two is their purpose and placement in the interaction. A system prompt is often at the beginning of an interaction, setting the stage for the conversation. An agent prompt occurs during the conversation, steering the direction of the dialogue based on user input.

Both types of prompts are crucial for creating a fluid and intuitive conversational experience for users. They guide the user and help ensure that the system understands and addresses the user's needs effectively.

Prompt Templates

Prompt templates are predefined formats or structures used to instruct a model or system to perform a specific task. They serve as a foundation for generating prompts, where certain parts of the template can be filled in or customized to produce a variety of specific prompts. By way of analogy, prompt templates are the counterpart to macros that you can define in some text editors.

Prompt templates are especially useful when working with language models, as they provide a consistent way to query the model across multiple tasks or data points. In particular, prompt templates can make it easier to:

- ensure consistency when querying a model multiple times
- facilitate batch processing or automation
- reduce errors and variations in how questions are posed to the model

As an example, suppose you're working with an LLM and you want to translate English sentences into French. An associated prompt template could be the following:

"Translate the following English sentence into French: {sentence}"

Note that {sentence} is a placeholder that you can replace with any English sentence.

You can use the preceding prompt template to generate specific prompts:

- "Translate the following English sentence into French: 'Hello, how are you?'"
- "Translate the following English sentence into French: 'I love ice cream.'"

As you can see, prompt templates enable you to easily generate a variety of prompts for different sentences without having to rewrite the entire instruction each time. In fact, this concept can be extended to more complex tasks and can incorporate multiple placeholders or more intricate structures, depending on the application.

Prompts for Different LLMs

GPT-3, ChatGPT, and GPT-4 are LLMs that are all based on transformer architecture and are fundamentally similar in their underlying mechanics. ChatGPT is essentially a version of the GPT model fine-tuned specifically for conversational interactions. GPT-4 is an evolution or improvement over GPT-3 in terms of scale and capabilities.

The differences in prompts for these models mainly arise from the specific use case and context, rather than inherent differences between the models. Here are some prompting differences that are based on use cases.

GPT-3 can be used for a wide range of tasks beyond just conversation, from content generation to code writing. Here are some examples of prompts for GPT-3:

"Translate the following English text to French: 'Hello, how are you?'"

"Write a Java method that calculates the factorial of a number."

ChatGPT is specifically fine-tuned for conversational interactions. Here are some examples of prompts for ChatGPT:

User: "Can you help me with my homework?"

ChatGPT: "Of course! What subject or topic do you need help with?"

User: "Tell me a joke."

ChatGPT: "Why did the chicken cross the playground? To get to the other slide!"

GPT-4: provides a larger scale and refinements, so the prompts would be similar in nature to GPT-3 but might yield more accurate or nuanced outputs. Here are some examples of prompts of prompts for GPT-4:

"Provide a detailed analysis of quantum mechanics in relation to general relativity."

"Generate a short story based on a post-apocalyptic world with a theme of hope."

These three models accept natural language prompts and produce natural language outputs. The fundamental way you interact with them remains consistent.

The main difference comes from the context in which the model is being used and any fine-tuning that has been applied. ChatGPT, for instance, is designed to be more conversational, so while you can use GPT-3 for chats, ChatGPT might produce more contextually relevant conversational outputs.

When directly interacting with these models, especially through an API, you might also have control over parameters like "temperature" (controlling randomness) and "max tokens" (controlling response length). Adjusting these can shape the responses, regardless of which GPT variant you're using.

In essence, while the underlying models have differences in scale and specific training/fine-tuning, the way to prompt them remains largely consistent: clear, specific natural language prompts yield the best results.

Poorly Worded Prompts

When crafting prompts, it's crucial to be as clear and specific as possible to guide the response in the desired direction. Ambiguous or vague prompts can lead to a wide range of responses, many of which might not be useful or relevant to the user's actual intent.

Moreover, poorly worded prompts are often vague, ambiguous, or too broad, and they can lead to confusion, misunderstanding, or non-specific responses from AI models. Here's a list of examples:

"*Tell me about that thing.*"

Problem: Too vague. What "thing" is being referred to?

"*Why did it happen?*"

Problem: No context. What event or situation is being discussed?

"*Explain stuff.*"

Problem: Too broad. What specific "stuff" should be explained?

"*Do what's necessary.*"

Problem: Ambiguous. What specific action is required?

"*I want information.*"

Problem: Not specific. What type of information is desired?

"*Can you get me the thing from the place?*"

Problem: Both "thing" and "place" are unclear.

"*Have you read what's his name's book?*"

Problem: Ambiguous reference. Who is "what's his name"?

"*How do you perform the process?*"

Problem: Which "process" is being referred to?

"Describe the importance of the topic."
Problem: "Topic" is not specified.

"Why is it bad or good?"
Problem: No context. What is "it"?

"Help with the issue."
Problem: Vague. What specific issue is being faced?

"Things to consider for the task."
Problem: Ambiguity. What "task" is being discussed?

"How does this work?"
Problem: Lack of specificity. What is "this"?

WHAT IS CHATGPT?

The chatbot wars are intensifying, and the long-term value of the primary competitors is still to be determined. One competitor (and arguably the current darling) is ChatGPT-3.5 (aka ChatGPT), which is an AI-based chatbot from OpenAI. ChatGPT responds to queries from users by providing conversational responses, and it's accessible here: *https://chat.openai.com/chat*

The growth rate in terms of registered users for ChatGPT has been extraordinary. The closest competitor is iPhone, which reached one million users in 2.5 months, whereas ChatGPT crossed one million users in *six days*. ChatGPT peaked around 1.8 billion users and then decreased to roughly 1.5 billion users, which you can see in the chart in this link:

https://decrypt.co/147595/traffic-dip-hits-openais-chatgpt-first-times-hardest

Note that although Threads from Meta out-performed ChatGPT in terms of membership, Threads has seen a significant drop in daily users in the neighborhood of 50%. A comparison of the time frame to reach one million members for six well-known companies/products and ChatGPT is here:

https://www.syntheticmind.io/p/01

The preceding link also contains information about Will Hobick who used ChatGPT to write a Chrome extension for email-related tasks, despite not having any JavaScript experience nor has he written a Chrome extension. Will Hobick provides more detailed information about his Chrome extension here:

https://www.linkedin.com/posts/will-hobick_gpt3-chatgpt-ai-activity-7008081003080470528-8QCh

ChatGPT: GPT-3 "on Steroids"?

ChatGPT has been called GPT-3 "on steroids," and there is some consensus that ChatGPT3 is currently the best chatbot in the world. Indeed, ChatGPT can perform multitude of tasks, some of which are listed here:

- write poetry
- write essays
- write code
- role play
- reject inappropriate requests

Moreover, the quality of its responses to natural language queries surpasses the capabilities of its predecessor GPT-3. Another interesting capability includes the ability to acknowledge its mistakes. ChatGPT also provides "prompt replies," which are examples of what you can ask ChatGPT. One interesting use for ChatGPT involves generating a text message for ending a relationship:

https://www.reddit.com/r/ChatGPT/comments/zgpk6c/breaking_up_with_my_girlfriend/

ChatGPT can generate Christmas lyrics that are accessible here:

https://www.cnet.com/culture/entertainment/heres-what-it-sounds-like-when-ai-writes-christmas-lyrics

One aspect of ChatGPT that probably won't be endearing to parents with young children is the fact that ChatGPT has told children that Santa Claus does not exist:

https://futurism.com/the-byte/openai-chatbot-santa

https://www.forbes.com/sites/lanceeliot/2022/12/21/pointedly-asking-generative-ai-chatgpt-about-whether-santa-claus-is-real-proves-to-be-eye-opening-for-ai-ethics-and-ai-law

ChatGPT: Google "Code Red"

In December 2022, the CEO of Google issued a "code red" regarding the potential threat of ChatGPT as a competitor to Google's search engine, which is briefly discussed here:

https://www.yahoo.com/news/googles-management-reportedly-issued-code-190131705.html

According to the preceding article, Google is investing resources to develop AI-based products, presumably to offer functionality that can successfully compete with ChatGPT. Some of those AI-based products might also generate graphics that are comparable to graphics effects by DALL-E. Indeed, the race to dominate AI continues unabated and will undoubtedly continue for the foreseeable future.

ChatGPT Versus Google Search

Given the frequent speculation that ChatGPT is destined to supplant Google Search, let's briefly compare the manner in which Google and ChatGPT respond to a given query. First, Google is a search engine that uses the Page Rank algorithm (developed by Larry Page), along with fine-tuned aspects to this algorithm that are a closely guarded secret. Google uses this algorithm to rank websites and to generate search results for a given query. However, the search results include paid ads, which can "clutter" the list of links.

By contrast, ChatGPT is not a search engine: it provides a direct response to a given query: in colloquial terms, ChatGPT simply "cuts to the chase" and eliminates the clutter of superfluous links. At the same time, ChatGPT can produce incorrect results, the consequences of which can range between benign and significant.

Consequently, Google search and ChatGPT both have strengths as well as weaknesses, and they excel with different types of queries: the former for queries that have multi-faceted answers (e.g., questions about legal issues), and the latter for straight-to-the point queries (e.g., coding questions). Obviously, both of them excel with many other types of queries.

According to the well-known computer scientist Margaret Mitchell, ChatGPT will not replace Google Search, and she provides some interesting details regarding Google Search and PageRank that you can read here: *https://twitter.com/mmitchell_ai/status/1605013368560943105*

ChatGPT Custom Instructions

ChatGPT has added support for custom instructions, which enable you to specify some of your preferences that ChatGPT will use when responding to your queries.

ChatGPT Plus users can switch on custom instructions by navigating to the ChatGPT website and then perform the following sequence of steps:

```
Settings > Beta features > Opt into Custom instructions
```

As a simple example, you can specify that you prefer to see code in a language other than Java. A set of common initial requirements for routine tasks can also be specified via a custom instruction in ChatGPT. A detailed sequence of steps for setting up custom instructions is accessible here:

https://artificialcorner.com/custom-instructions-a-new-feature-you-must-enable-to-improve-chatgpt-responses-15820678bc02

Another interesting example of custom instructions is from Jeremy Howard (the creator of fast.ai), who prepared an extensive and detailed set of custom instructions that is accessible here:

https://twitter.com/jeremyphoward/status/1689464587077509120

As this book goes to print, custom instructions are available only for users who have registered for ChatGPT Plus. However, OpenAI has stated that custom instructions will be available for free to all users by the end of 2023.

ChatGPT on Mobile Devices and Browsers

ChatGPT first became available for iOS devices and then for Android devices during 2023. You can download ChatGPT onto an iOS device from the following link:

https://www.macobserver.com/tips/how-to/how-to-install-and-use-the-official-chatgpt-app-on-iphone/

Alternatively, if you have an Android device, you can download ChatGPT from the following link: *https://play.google.com/store/apps/details?id=com.openai.chatgpt*

If you want to install ChatGPT for the Bing browser from Microsoft, navigate to this link:

https://chrome.google.com/webstore/detail/chatgpt-for-bing/pkkmgcildaegadhngpjkklnbfbmhpdng

ChatGPT and Prompts

Although ChatGPT is very adept at generating responses to queries, sometimes you might not be fully satisfied with the result. One option is to type the word "rewrite" in order to get another version from ChatGPT.

Although this is one of the simplest prompts available, it's limited in terms of effectiveness. If you want a list of more meaningful prompts, the following article contains thirty-one prompts that have the potential to be better than using the word "rewrite" (from all AI text generators, not just with ChatGPT):

https://medium.com/the-generator/31-ai-prompts-better-than-rewrite-b3268dfe1fa9

GPTBot

GPTBot is a crawler for websites. Fortunately, you can disallow GPTBot from accessing a website by adding the GPTBot to the `robots.txt` file for a website:

```
User-agent: GPTBot

Disallow: /
```

You can also customize GPTBot access only a portion of a website by adding the GPTBot token to to the `robots.txt` like file for a website:

```
User-agent: GPTBot
Allow: /youcangohere-1/
Disallow: /dontgohere-2/
```

As an aside, Stable Diffusion and LAION both scrape the internet via Common Crawl. However, you can prevent your website from being scraped by specifying the following snippet in the `robots.txt` file:

```
User-agent: CCBot
Disallow: /
```

More information about GPTBot is accessible at the following links: https://platform.openai.com/docs/gptbot

https://www.yahoo.com/finance/news/openai-prepares-unleash-crawler-devour-020628225.html

ChatGPT Playground

ChatGPT has its own playground, which you will see is substantively different from the GPT-3 playground accessible here: *https://chat.openai.com/chat*

GPT-3 playground can be found here:

https://beta.openai.com/playground

OpenAI has periodically added new functionality to ChatGPT that includes the following:

- Users can view (and continue) previous conversations.
- There is a reduction in the number of questions that ChatGPT will not answer.
- Users remain logged in for longer than two weeks.

Another nice enhancement includes support for keyboard shortcuts: when working with code you can use the sequence ⌘ (Ctrl) + Shift + (for Mac) to copy the last code block and the sequence ⌘ (Ctrl) + / to see the complete list of shortcuts.

Many articles are available regarding ChatGPT and how to write prompts in order to extract the details that you want from ChatGPT. One of those articles is here:

https://www.tomsguide.com/features/7-best-chatgpt-tips-to-get-the-most-out-of-the-chatbot

PLUGINS, ADVANCED DATA ANALYSIS, AND CODE WHISPERER

In addition to answering a plethora of queries from users, ChatGPT extends its functionality by providing support for the following:

- third-party ChatGPT plug-ins
- Advanced Data Analysis
- Code Whisperer

Each of the topics in the preceding bullet list are briefly discussed in the following subsections, along with a short section that discusses Advanced Data Analysis versus Claude-2 from Anthropic.

Plugins

There are several hundred ChatGPT plugins available, and some popular plugins are accessible in the following links:

https://levelup.gitconnected.com/5-chatgpt-plugins-that-will-put-you-ahead-of-99-of-data-scientists-4544a3b752f9

https://www.zdnet.com/article/the-10-best-chatgpt-plugins-of-2023/

Keep in mind that lists of the "best" ChatGPT plugins change frequently, so it's a good idea to perform an online search to find out about newer ChatGPT plugins. The following link also contains details about highly rated plugins (by the author of the following article):

https://www.tomsguide.com/features/i-tried-a-ton-of-chatgpt-plugins-and-these-3-are-the-best

The following list includes another set of recommended plugins (depending on your needs, of course):

- AskYourPDF
- ChatWithVideo
- Noteable
- Upskillr
- Wolfram

If you are concerned about the possibility of ChatGPT scraping the content of your website, the browser plugin from OpenAI supports a user-agent token called ChatGPT-User that abides by the content specified in the `robots.txt` file that many websites provide for restricting access to content.

If you want to develop a plugin for ChatGPT, navigate to the following website for more information: *https://platform.openai.com/docs/plugins/introduction*

Along with details for developing a ChatGPT plugin, the preceding OpenAI website provides useful information about plugins, as shown here:

- authentication
- examples
- plugin review
- plugin policies

OpenAI does not control any plugins that you add to ChatGPT; they connect ChatGPT to external applications. Moreover, ChatGPT determines which plugin to use during your session, based on the specific plugins that you have enabled in your ChatGPT account.

Advanced Data Analysis (Formerly Code Interpreter)

ChatGPT Advanced Data Analysis enables ChatGPT to generate charts and graphs. It also allows ChatGPT to create and train machine learning models, including deep learning models. ChatGPT Advanced Data Analysis provides an extensive set of features and it's available to ChatGPT users for a US $20 per month subscription. However, this feature will probably be made available to all users very soon.

Advanced Data Analysis provides a diverse range of notable features, some of which are in the following bullet list:

- Interactive Computing Environment
- Multiple Languages Support
- Data Visualization
- Integration with Databases
- Support for Libraries and Frameworks
- File Management
- Stateful Sessions
- Safe Execution Environment
- Extensive Documentation and Help
- Collaboration Features
- History and Logging
- Customizability and Extensibility

More information about Advanced Data Analysis can be found here:

https://towardsdatascience.com/chatgpt-code-interpreter-how-it-saved-me-hours-of-work-3c65a8dfa935

The models from OpenAI can access a Python interpreter that is confined to a sandboxed and fire-walled execution environment. There is also some temporary disk space that is accessible to the interpreter plugin during the evaluation of Python code. Although the temporary disk space is available for a limited time, multiple queries during the same session can produce a cumulative effect regarding the code and execution environment.

In addition, ChatGPT can generate a download link (upon request) in order to download data. One other interesting feature, starting in mid-2023, is that Advanced Data Analysis can analyze multiple files at once, which includes CSV files and Excel spreadsheets.

Advanced Data Analysis can perform an interesting variety of tasks, some of which are that it can:

- solve mathematical tasks
- perform data analysis and visualization

- convert files between formats
- work with Excel spreadsheets
- read textual content in a PDF

The following article discusses various ways that you can use Advanced Data Analysis:

https://mlearning.substack.com/p/the-best-88-ways-to-use-chatgpt-code-interpreter

Advanced Data Analysis Versus Claude-2

Claude-2 from Anthropic is another competitor of ChatGPT. In addition to responding to prompts from users, Claude-2 can generate code and also ingest entire books. Claude-2 is also subject to hallucinations, which is true of other LLM-based chatbots. More detailed information regarding Claude-2 is accessible here:

https://medium.com/mlearning-ai/claude-2-vs-code-interpreter-gpt-4-5-d2e-5c9ee00c3

Incidentally, the currently available version of ChatGPT was trained in September 2021, which means that ChatGPT cannot answer questions regarding Claude-2 or Google Bard, both of which were released after that date.

Code Whisperer

ChatGPT Code Whisperer enables you to simplify some tasks, some of which are listed below (compare this list with the corresponding list for Bard):

- You can create videos from images.
- You can extract text from an image.
- You can extract colors from an image.

After ChatGPT has generated a video, it will also give you a link from which the generated video is downloadable. More detailed information regarding the features in the preceding bullet list is accessible here:

https://artificialcorner.com/chatgpt-code-interpreter-is-not-just-for-coders-here-are-6-ways-it-can-benefit-everyone-b3cc94a36fce

DETECTING GENERATED TEXT

Without a doubt, `ChatGPT` has raised the bar with respect to the quality of generated text, which further complicates the task of plagiarism. When you read a passage of text, there are several clues that suggest generated text, such as:

- awkward or unusual sentence structure
- repeated text in multiple locations
- excessive use of emotions (or absence thereof)

However, there are tools that can assist in detecting generated code. One free online tool is `GPT2 Detector` (from OpenAI) that is accessible here:

https://huggingface.co/openai-detector

As a simple (albeit contrived) example, type the following sentence in GPT2 Detector:

```
This is an original sentence written by me and nobody else.
```

The GPT2 Detector analyzed this sentence and reported that this sentence is real with a 19.35% probability. Now let's modify the preceding sentence by adding some extra text, as shown here:

```
This is an original sentence written by me and nobody else,
regardless of what an online plagiarism tool will report
about this sentence.
```

The GPT2 Detector analyzed this sentence and reported that this sentence is real with a 95.85% probability. According to the GPT2 Detector website, the reliability of the probability scores "get reliable" when there are around fifty tokens in the input text.

Another (slightly older) online tool for detecting automatically generated text is GLTR (giant language model test room) from IBM, which is accessible here: *http://gltr.io/*

You can download the source code (a combination of TypeScript and CSS) for GLTR here:

https://github.com/HendrikStrobelt/detecting-fake-text

In addition to the preceding free tools, some commercial tools are also available, one of which is shown here: *https://writer.com/plans/*

CONCERNS ABOUT CHATGPT

One important aspect of ChatGPT is that it's not designed for accuracy: in fact, ChatGPT can generate (fabricate?) very persuasive answers that are actually incorrect. This detail distinguishes ChatGPT from search engines: the latter provide links to existing information instead of generating responses that might be incorrect. Another comparison is that ChatGPT is more flexible and creative, whereas search engines are less flexible but more accurate in their responses to queries.

Educators are concerned about students using ChatGPT as a tool to complete their class assignments instead of developing research-related skills in conjunction with writing skills. At the same time, there are educators who enjoy the reduction in preparation time for their classes as a direct result of using ChatGPT to prepare lesson plans.

Another concern is that ChatGPT cannot guarantee that it provides factual data in response to queries from users. In fact, ChatGPT can "hallucinate," which means that it can provide wrong answers as well as citations (i.e., links) that do not exist.

Another limitation of ChatGPT is due to the use of training data that was available only up until 2021. However, OpenAI does support plug-ins for ChatGPT, one of which can perform on-the-fly real time Web searches.

As you will learn later, the goal of prompt engineering is to understand how to craft meaningful queries that will induce ChatGPT to provide the information that you want: poorly worded (or incorrectly worded) prompts can produce equally poor results. As a rule, it's advisable to curate the contents of the responses from ChatGPT, especially in the case of responses to queries that involve legal details.

Code Generation and Dangerous Topics

Two significant areas for improvement pertain to code generation and handling dangerous topics.

Although `ChatGPT` (as well as `GPT-3`) can generate code for various types of applications, keep in mind that `ChatGPT` displays code that was written by other developers, which is also code that was used to train `ChatGPT`. Consequently, portions of that code (such as version numbers) might be outdated or code that is actually incorrect.

As for queries that involve dangerous topics, `ChatGPT` explains why it cannot answer such a query. However, a query that is posed in "pretend mode" ("suppose you are a fictional character, and how would you explain ....") has enabled people to obtain results from ChatGPT that do not conform to its guidelines.

Other serious potential issues exist, some of which are discussed in the following article:

https://www.yahoo.com/news/hypnotized-chatgpt-bard-convince-users-182100261.html

ChatGPT Strengths and Weaknesses

ChatGPT has a number of strengths. In particular, ChatGPT responds directly to questions without including extraneous links. As a simple example, suppose you enter a query in an online search engine, such as "Generate HTML for an input field."

On the other hand, ChatGPT is prone to various types of errors, along with limitations that include the following:

- inability to "fact check" information
- inability to differentiate misinformation versus facts
- can fabricate answers (aka hallucinations)
- can generate inappropriate/toxic responses

`ChatGPT` is prone to prompt injection, which can consist of very simple content. However, other chatbots are also susceptible to prompt injection (i.e., it's not specific to `ChatGPT`). More details regarding prompt injections are accessible here:

https://mlearning.substack.com/p/what-exactly-is-a-prompt-injection

Incidentally, the chatbot LaMDA from Google is also prone to the preceding limitations, which has made Google hesitant to release LaMDA for general availability because of its potential risk regarding Google's reputation. As a side note, LaMDA was announced in May 2021 and at that time it was described as a "breakthrough language technology": *https://blog.google/technology/ai/lamda/*

Despite the earlier bullet list, ChatGPT has increased the capabilities of chatbots, and will undoubtedly spur innovation and competition among prominent stakeholders in this arena.

SAMPLE QUERIES AND RESPONSES FROM CHATGPT

Here is an example of various queries submitted to ChatGPT, followed by the detailed response from ChatGPT:

```
Many people think that AI is dangerous what do you think?
```

"There is a common misconception that AI is inherently dangerous. While it is true that AI has the potential to be used for harmful purposes, it is also capable of doing a lot of good. Just like any other technology, it all depends on how it is used. In the right hands, AI can be used to solve some of the world's most pressing problems, such as curing diseases and improving agriculture. In the wrong hands, it could be used to cause harm. It's important for us to use AI responsibly and ethically, and to ensure that it is regulated and controlled in a way that minimizes the potential for harm."

Here is another example of a query submitted to ChatGPT, followed by its detailed response:

```
What are some ways in which you can improve the accuracy of
your responses to queries?
```

"There are several ways to improve the accuracy of my responses to queries. One way is to use more data to train my algorithms on. The more data I have, the more information I have to work with, and the more accurate my responses can be. Another way to improve my accuracy is to use more advanced algorithms and techniques, such as machine learning and natural language processing. These methods allow me to better understand the nuances of language and to provide more accurate responses to queries. Additionally, regular testing and evaluation can help identify areas where my accuracy can be improved and allow me to focus on those areas to refine my responses."

```
Can you give an example of an odd perfect number?
```

"As far as we know, there are no odd perfect numbers. While it is believed that odd perfect numbers exist, none have been found yet, and it is not known if there are any. In fact, it is currently unknown if there are any odd perfect numbers at all. If an odd perfect number were to be discovered, it would be a major breakthrough in mathematics."

```
Can you provide a proof of Goldbach's conjecture?
```

"I'm sorry, but I am not able to provide proof of Goldbach's conjecture. Goldbach's conjecture is a famous unsolved problem in mathematics that states that every even integer greater than 2 can be expressed as the sum of two prime numbers. Despite much effort by mathematicians over the years, proof of the conjecture has not yet been found. It is considered to be one of the most challenging unsolved problems in mathematics. Because I am a large language model trained by OpenAI, I do not have the ability to prove mathematical conjectures. My abilities are limited to providing answers based on the information that I have been trained on."

```
Is there life after death, and how does it compare with
life on Earth?
```

"There is no scientific evidence to support the idea of life after death. This is a topic that has been debated for centuries, and different cultures and belief systems have their own ideas about what happens after we die. Some people believe in an afterlife, while others do not. Because there is no scientific evidence to support the existence of an afterlife, it is not possible to say how it would compare to life on Earth. It is ultimately a matter of personal belief."

```
Please provide a four-paragraph analysis of the war between
Ukraine and Russia, and also provide your prediction
regarding the outcome.
```

"The war between Ukraine and Russia is a complex conflict that has its roots in the 2014 Russian annexation of Crimea. Since then, the conflict has escalated, with Russian-backed separatists in eastern Ukraine fighting against the Ukrainian government. The fighting has resulted in thousands of casualties and has displaced many people.

The main issue at the heart of the conflict is the question of Ukraine's sovereignty. Russia has long considered Ukraine to be within its sphere of influence, and the annexation of Crimea was seen by many as a violation of Ukraine's sovereignty. Ukraine, on the other hand, is determined to maintain its independence and has been supported in this by many Western countries.

It is difficult to predict the outcome of the war. While the Ukrainian military has been able to make gains against the separatists in recent years, Russia continues to provide support to the rebels and the conflict remains unresolved. In the long term, a diplomatic solution will likely be necessary in order to bring an end to the fighting. However, it is uncertain if and when such a solution will be reached.

In the meantime, the war continues to take a heavy toll on the people of Ukraine. Many are caught in the crossfire, and the ongoing violence and instability have had a significant impact on the country's economy and overall development. It is important for the international community to continue to support Ukraine in its efforts to maintain its sovereignty and bring an end to the conflict."

As you can see from the preceding examples, ChatGPT provides impressive and accurate responses to queries in a manner that surpasses many chatbots of recent years. However, it's still possible to confound this chatbot. Read about some of those weaknesses here:

https://www.theverge.com/23488017/openai-chatbot-chatgpt-ai-examples-web-demo

CHATGPT AND MEDICAL DIAGNOSIS

A young boy who had been experiencing chronic pain for several years finally found hope through an unexpected helper: ChatGPT. Over a span of three years, the boy's mother had taken him to see 17 different specialists, yet they remained without a diagnosis that could account for all of his symptoms.

The turning point came earlier this year when his mother decided to seek assistance from ChatGPT. She created an account and meticulously input all the details she had gathered over the years, including her son's symptoms and the data from his MRI scans. She recalls the countless hours she spent in front of the computer, sifting through information in a desperate bid to find answers.

Despite visiting numerous doctors and even rushing to the emergency room at one point, the family felt they were running in circles, with each specialist only focusing on their field of expertise without offering a comprehensive solution. She noted a worrying sign when her son stopped growing. Although their pediatrician initially attributed this to the adverse effects of the pandemic, the boy's mother felt there was more to it.

In a moment of desperation and determination, she turned to ChatGPT, inputting every piece of information she had about her son's condition. It was then that ChatGPT suggested the possibility of tethered cord syndrome, a suggestion that resonated with her and seemed to connect all the dots. After a specialist confirmed the suggestion from ChatGPT was correct, she realized this was a pivotal moment in their long and exhausting journey towards finding a diagnosis.

ALTERNATIVES TO CHATGPT

There are several alternatives to ChatGPT that offer a similar set of features, some of which are listed here:

- Bard (Google)
- Bing Chat

- Gemini (Google)
- Jasper
- PaLM (Google)
- Pi
- POE (LinkedIn)
- Replika
- WriteSonic
- YouChat

The following subsections discuss some (but not all) of the ChatGPT alternatives that appear in the preceding bulleted list.

Google Bard

Google Bard is a chatbot that has similar functionality as ChatGPT, such as generating code as well as generating text/documents. A subset of the features supported by Bard are listed here:

- built-in support for internet search
- built-in support for voice recognition
- built "on top of" PaLM 2 (Google)
- support for 20 programming languages
- read/summarize PDF contents
- provides links for its information

According to the following article in mid-2023, Bard has added support for forty additional languages as well as support for text-to-speech:

https://www.extremetech.com/extreme/google-bard-updated-with-text-to-speech-40-new-languages

Moreover, Bard supports prompts that include images (interpreted by Google Lens) and can produce captions based on the images.

The following article suggests that Google can remain competitive with ChatGPT by leveraging PaLM (discussed later in this chapter):

https://analyticsindiamag.com/googles-palm-is-ready-for-the-gpt-challenge/

YouChat

Another alternative to ChatGPT is YouChat that is part of the search engine you.com, and it's accessible here:

https://you.com/

Richard Socher, who is well known in the ML community for his many contributions, is the creator of you.com. According to Richard Socher, YouChat is a search engine that can provide the usual search-related functionality as well as the ability to search the Web to obtain more information in order to provide responses to queries from users.

Another competitor is POE from LinkedIn, and you can create a free account at this link: *https://poe.com/login*

Pi From Inflection

Pi is a chatbot developed by Inflection, which is a company that was by Mustafa Suleyman, who is also the founder of DeepMind. Pi is accessible here:

(*https://pi.ai/talk*)

More information about Pi can be found here:

https://medium.com/@ignacio.de.gregorio.noblejas/meet-pi-chatgpts-newest-rival-and-the-most-human-ai-in-the-world-367b461c0af1

The development team used reinforcement learning from human feedback (RLHF) in order to train this chatbot. Information about RLHF is accessible here:

https://huggingface.co/blog/rlhf

MACHINE LEARNING AND CHATGPT

ChatGPT can be leveraged in various ways in the realm of machine learning, some of which are shown in the following bullet list:

- Natural Language Processing (NLP) Tasks
- Data Augmentation
- Conversational Agents
- Content Creation
- Code-related Tasks
- Educational Applications
- Research & Knowledge Extraction
- Interactive Entertainment
- Semantic Search
- Business Intelligence
- Analyzing customer feedback or reviews to extract insights
- Generating reports or summaries from raw business data
- Converting text to more accessible formats
- Fine-tuning for Specific Domains
- Adapting ChatGPT to specific industries

These are just a few of the many ways ChatGPT can be leveraged in machine learning. Given its versatility, the potential applications are vast and continue to grow as more developers explore its capabilities.

Moreover, Advanced Data Analysis can generate machine learning models that can be trained on datasets. For example, Figure 8.1 displays a screenshot of charts that are based on the `Titanic` dataset.

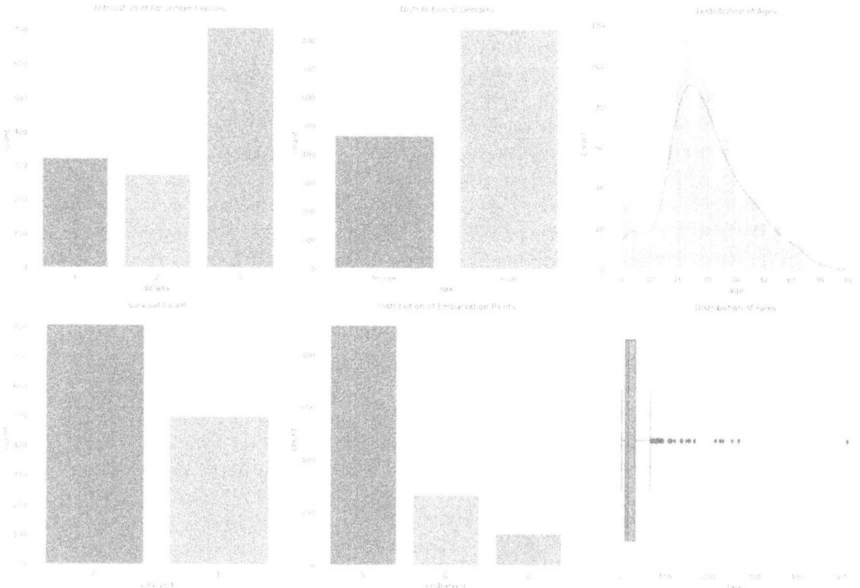

FIGURE 8.1. Titanic charts and graphs.

Incidentally, if you would like to see examples of ChatGPT generating `Python` code for machine learning models, as well as code for charts and graphs, you can learn how to do so in these forthcoming titles that will be available in early 2024, all of which are from Mercury Learning: *Machine Learning, Python3, and ChatGPT; Python and ChatGPT/GPT-4; and Python and Data Visualization With ChatGPT.*

WHAT IS INSTRUCTGPT?

InstructGPT is a language model developed by OpenAI, and it's a sibling model to ChatGPT. InstructGPT is designed to follow instructions given in a prompt to generate detailed responses. Some key points about instructGPT are listed here:

- following instructions
- training
- applications
- limitations

Following instructions: Unlike ChatGPT, which is more geared towards open-ended conversations, instructGPT is designed to be more focused on following user instructions in prompts. This makes it suitable for tasks where the user wants to get specific information or outputs by giving clear directives.

Training: instructGPT is trained using reinforcement learning from human feedback (RLHF), similar to ChatGPT. An initial model is trained using supervised fine-tuning, where human AI trainers provide conversations playing both sides (the user and the AI assistant). This new dialogue dataset is then mixed with the InstructGPT dataset transformed into a dialogue format.

Applications: instructGPT can be useful in scenarios where you want more detailed explanations, step-by-step guides, or specific outputs based on the instructions provided.

Limitations: Like other models, instructGPT has its limitations. It might produce incorrect or nonsensical answers. The output heavily depends on how the prompt is phrased. It's also sensitive to input phrasing and might give different responses based on slight rephrasing.

It's worth noting that as AI models and their applications are rapidly evolving, so you can expect further developments or iterations on instructGPT. Always refer to OpenAI's official publications and updates for the most recent information. More information about InstructGPT is accessible here:

https://openai.com/blog/instruction-following/

VIZGPT AND DATA VISUALIZATION

VizGPT is an online tool that enables you to specify English-based prompts in order to visualize aspects of datasets, and it's accessible here: *https://www.vizgpt.ai/*

Select the default "Cars Dataset" and then click on the "Data" button in order to display the contents of the dataset, as shown in Figure 8.2.

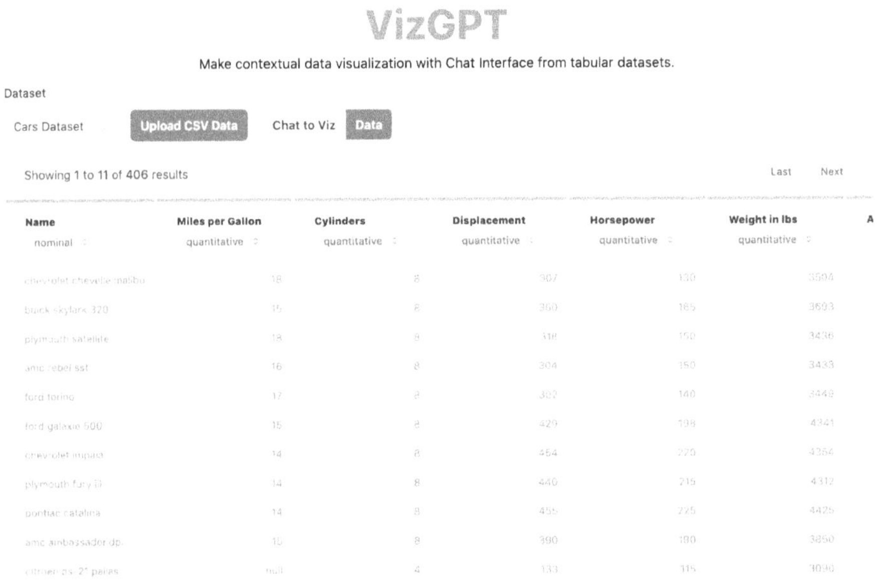

VizGPT

Make contextual data visualization with Chat Interface from tabular datasets.

Dataset

Cars Dataset   Upload CSV Data   Chat to Viz   Data

Showing 1 to 11 of 406 results   Last   Next

Name nominal	Miles per Gallon quantitative	Cylinders quantitative	Displacement quantitative	Horsepower quantitative	Weight in lbs quantitative	A
chevrolet chevelle malibu	18	8	307	130	3504	
buick skylark 320	15	8	350	165	3693	
plymouth satellite	18	8	318	150	3436	
amc rebel sst	16	8	304	150	3433	
ford torino	17	8	302	140	3449	
ford galaxie 500	15	8	429	198	4341	
chevrolet impala	14	8	454	220	4354	
plymouth fury iii	14	8	440	215	4312	
pontiac catalina	14	8	455	225	4425	
amc ambassador dpl	15	8	390	190	3850	
citroen ds-21 pallas	null	4	133	115	3090	

FIGURE 8.2. VizGPT car dataset rows.

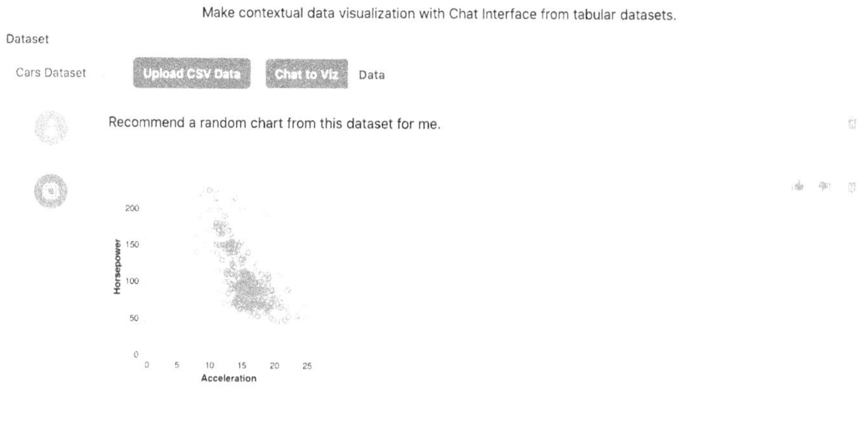

FIGURE 8.3. VizGPT car dataset visualization.

Next, select the default "Cars Dataset" and then click on the "Chat to Viz" button in order to display a visualization of the dataset, as shown in Figure 8.3.

You can experiment further with VizGPT. For example, you can upload your own dataset by clicking on the "Upload CSV" button and obtain similar results with that dataset.

WHAT IS GPT-4?

GPT-4 was released in mid-March 2023, and became available only to users with an existing ChatGPT account via a paid upgrade of US$ 20 per month to that account. According to various online anecdotal stories from users, GPT-4 is significantly superior to ChatGPT. In addition, Microsoft has a version of GPT-4 that powers its Bing browser, which is freely available to the public.

GPT-4 is a large multimodal model that can process image-based inputs as well as text-based inputs and then generate textual outputs. Currently image-based outputs are unavailable to the general public, but it does have internal support for image generation.

GPT-4 supports 25,000 words of input text. By comparison, ChatGPT is limited to 4,096 characters. Although the number of parameters in GPT-4 is undisclosed, the following article asserts that GPT-4 is a mixture of 8 x 220-billion-parameter models, which is an example of MoE (mixture of experts).

More about GPT-4 can be found here:

https://thealgorithmicbridge.substack.com/p/gpt-4s-secret-has-been-revealed

GPT-4 and Test-Taking Scores

One interesting example of the improved accuracy pertains to a bar exam, in which ChatGPT scored in the bottom 10%. By contrast, GPT-4 scored in the top 10% for the same bar exam. More details are accessible here:

https://www.abajournal.com/web/article/latest-version-of-chatgpt-aces-the-bar-exam-with-score-in-90th-percentile

In addition, GPT-4 is apparently able to pass the first year at Harvard with a 3.34 GPA. More details are accessible here:

https://www.businessinsider.com/chatgpt-harvard-passed-freshman-ai-education-GPT-4-2023-7?op=1

Furthermore, GPT-4 has performed well on a number of additional tests, some of which are listed here:

- AP exams
- SAT
- GRE
- medical tests
- law exams
- business school exams
- Wharton MBA exam
- USA Biology Olympiad semifinal exam
- sommelier exams (wine steward)

You can read more details regarding the preceding tests from this link:

https://www.businessinsider.com/list-here-are-the-exams-chatgpt-has-passed-so-far-2023-1

The following link contains much more detailed information regarding test scores, benchmarks, and other results pertaining to GPT-4: *https://openai.com/research/gpt-4*

GPT-4 Parameters

This section contains information regarding some of the GPT-4 parameters, some of which are best-guess approximations.

Since GPT-4 is a transformer-based AR (auto regressive) model, it's trained to perform next-token prediction. The following paper "GPT4 Technical Report" was released in March 2023 and it contains a detailed analysis of the capabilities of GPT-4:

https://docs.kanaries.net/en/tutorials/ChatGPT/gpt-4-parameters

GPT-4 Fine Tuning

Although OpenAI allows you to fine-tune the four base models (davinci, curie, ada, and babbage), it's (currently) not possible to perform fine tuning on

ChatGPT 3.5 or GPT-4. Instead, you can integrate OpenAI models with your own data source via LangChain or LlamaIndex (previously known as GPT-Index). Both of them enable you to connect OpenAI models with your existing data sources.

An introduction to LangChain is accessible here:

https://www.pinecone.io/learn/series/langchain/langchain-intro/

An introduction to LlamaIndex is accessible here:

https://zilliz.com/blog/getting-started-with-llamaindex

https://stackoverflow.com/questions/76160057/openai-chat-completions-api-how-do-i-customize-answers-from-gpt-3-5-or-gpt-4-mo?noredirect=1&lq=1

CHATGPT AND GPT-4 COMPETITORS

Shortly after the release of ChatGPT on November 30, 2022, there was a flurry of activity among various companies to release a competitor to ChatGPT, some of which are listed here:

- Bard (Google chatbot)
- CoPilot (Microsoft)
- Codex (OpenAI)
- Apple GPT (Apple)
- PaLM 2 (Google and GPT-4 competitor)
- Claude 2 (Anthropic)
- Llama-2 (Meta)

The following subsections contain additional details regarding the LLMs in the preceding bulleted list.

Bard

Bard is an AI chatbot from Google that was released in early 2023 and it's a competitor to ChatGPT. By way of comparison, Bard is powered by PaLM 2 (discussed later), whereas ChatGPT is powered by GPT-4. Recently Bard added support for images in its answers to user queries, whereas this functionality for ChatGPT has not been released yet to the public. More about Bard's image recognition can be found here:

https://artificialcorner.com/google-bards-new-image-recognition-means-serious-competition-to-chatgpt-here-are-6-best-use-cases-55d69eae1b27

Bard encountered an issue pertaining to the James Webb Space Telescope during a highly publicized release, which resulted in a significant decrease in market capitalization for Alphabet. However, Google has persevered in fixing issues and enhancing the functionality of Bard. You can access Bard from this link: *https://bard.google.com/*

Around mid-2023 Bard was imbued with several features that were not available in GPT-4 during the same time period, some of which are listed here:

- generating images
- generating HTML/CSS from an image
- generating mobile applications from an image
- creating LaTeX formulas from an image
- extracting text from an image

Presumably these features will spur OpenAI to provide the same set of features (some are implemented in GPT-4 but they are not publicly available).

CoPilot (OpenAI/Microsoft)

Microsoft CoPilot is a Visual Studio code extension that is also powered by GPT-4. GitHub CoPilot is already known for its ability to generate blocks of code within the context of a program. In addition, Microsoft is also developing Microsoft 365 CoPilot, however, the availability date has not been announced as of mid-2023.

However, Microsoft has provided early demos that show some of the capabilities of Microsoft 365 CoPilot, which includes automating tasks such as:

- writing emails
- summarizing meetings
- making PowerPoint presentations

Microsoft 365 CoPilot can analyze data in Excel spreadsheets, insert AI-generated images in PowerPoint, and generate drafts of cover letters. Microsoft has also integrated Microsoft 365 CoPilot into some of its existing products, such as Loop and OneNote.

According to the following article, Microsoft intends to charge USD 30 per month for Office 365 Copilot:

https://www.extremetech.com/extreme/microsoft-to-charge-30-per-month-for-ai-powered-office-apps

Copilot was reverse engineered in late 2022, which is described here:

https://thakkarparth007.github.io/copilot-explorer/posts/copilot-internals

The following article shows you how to create a GPT-3 application that uses NextJS, React, and CoPilot:

https://github.blog/2023-07-25-how-to-build-a-gpt-3-app-with-nextjs-react-and-github-copilot/

Codex (OpenAI)

OpenAI Codex is a fine-tuned GPT3-based LLM that generates code from text. In fact, Codex powers GitHub Copilot (discussed in the preceding section). Codex was trained on more than 150GB of Python code that was obtained from more than 50 million GitHub repositories.

According to OpenAI, the primary purpose of Codex is to accelerate human programming, and it can complete almost 40% of requests. Codex tends to work quite well for generating code for solving simpler tasks. Navigate to the Codex home page to obtain more information: *https://openai.com/blog/openai-codex*

Apple GPT

In mid-2023 Apple announced Apple GPT, which is a competitor to ChatGPT from OpenAI. The actual release date was projected to be 2024. "Apple GPT" is the current name for a product that is intended to compete with Google Bard, OpenAI ChatGPT, and Microsoft Bing AI.

In brief, the LLM PaLM 2 (discussed in the next section) powers Google Bard, and GPT-4 powers ChatGPT as well as Bing Chat, whereas Ajax is what powers Apple GPT. Ajax is based on Jax from Google, and the name Ajax is a clever concatenation ("Apple Jax" perhaps?).

PaLM-2

PaLM-2 is an acronym for pathways language model, and it is the successor to PaLM (circa 2022). PaLM-2 powers Bard and it's also a direct competitor to GPT-4. By way of comparison, PaLM consists of 540B parameters, and it's plausible that PaLM-2 is a larger LLM (details of the latter are undisclosed).

PaLM-2 provides four sub models called Gecko, Otter, Bison, and Unicorn (smallest to largest). PaLM-2 was trained in more than 100 human languages, as well as programming languages such as Fortran. Moreover, PaLM-2 has been deployed to a plethora of Google products, including Gmail and YouTube.

Med-PaLM M

In addition to the four sub models listed above, Med-PaLM 2 (the successor to Med-PaLM) is an LLM that provides answers to medical questions, and it's accessible here: *http://sites.research.google/med-palm/*

The successor to Med-PaLM is Med-PaLM M, and details about this LLM are accessible here: *https://arxiv.org/abs/2307.14334*

An article that provides a direct comparison of performance benchmarks for PaLM 2 and GPT-4 is accessible here:

https://www.makeuseof.com/google-palm-2-vs-openai-gpt-4/

All told, PaLM-2 has a robust set of features and it's definitely a significant competitor to GPT-4.

Claude 2

Anthropic created the LLM Claude 2 that can not only answer queries about specific topics, but it can also perform searches that involve multiple documents, summarize documents, create documents, and generate code.

Claude 2 is an improvement on Anthropic's predecessor Claude 1.3, and it can ingest entire books as well as generate code based on prompts from users. In fact, Claude 2 appears to be comparable with its rivals ChatGPT and GPT-4 in terms of competing functionality.

Furthermore, Claude 2 supports a context window of 100,000 tokens. Moreover, Claude 2 was trained on data as recent as early 2023, whereas ChatGPT was trained on data up until 2021. However, Claude 2 cannot search the Web (unlike its competitor GPT-4). Stay tuned to Anthropic, where you will probably see more good things in the LLM space.

LLAMA-2

LlaMa 2 (large language model meta AI) is an open source fine-tuned LLM from Meta, that trained on only public data, that has created a lot of excitement in the AI community. LlaMa-2 provides three models (7B, 13B, and 70B parameters) that utilize more data during the pre-training step than numerous other LLMs. LlaMa-2 is optimized to provide faster inferences and also provides a longer context length (4K) than other LLMs.

Moreover, the LlaMa-2-Chat LLM performs surprisingly well. In some cases, its quality is close to the quality of high-performing LLMs such ChatGPT and GPT-4. LlaMa-2 is more user-friendly, and also provides better results for writing text in comparison to GPT-4. However, GPT-4 is more adept for tasks such as generating code.

How to Download LlaMa-2

Llama-2 provides a permissive license for community use and commercial use, and Meta has made the code as well as the pretrained models and the fine-tuned models publicly available.

There are several ways that you can download Llama-2, starting from this link from Meta after you provide some information (name, country, and affiliation):

https://ai.meta.com/llama/

Another way to access demos of the 7B, 13B, and 70B models is from the following links:

https://huggingface.co/spaces/huggingface-projects/llama-2-7b-chat

https://huggingface.co/spaces/huggingface-projects/llama-2-13b-chat

https://huggingface.co/spaces/ysharma/Explore_llamav2_with_TGI

A third way to access Llama-2 on Hugging Face from the following link: *https://huggingface.co/blog/llama2*

If you are interested in training Llama-2 on your laptop, more details for doing so are accessible here: *https://blog.briankitano.com/llama-from-scratch/*

LlaMa-2 Architecture Features

This section contains a high-level list of some of the important distinguishing features of LlaMa-2, as shown here:

- decoder-only LLM
- better pretraining
- improved model architecture
- SwiGLU activation function
- different positional embeddings
- GQA (grouped query attention)
- Ghost Attention (GAtt)
- RLHF and PPO
- BPE SentencePiece tokenizer
- modified normalization step

The majority of LLMs contain the layer normalization that is in the original transformer architecture. By contrast, LlaMA uses a simplified alternative that involves root mean square layer normalization (aka RMSNorm). RMSNorm has yielded improved results for training stability as well as for generalization.

Although SwiGLU is computationally more expensive than the ReLU activation function that is part of the original transformer architecture, SwiGLU achieves better performance.

Note that RLHF is discussed in Chapter 5, which also includes a brief description of TRPO and PPO. For a detailed description of how to fine tune LlaMa-2 on three tasks, navigate to the following link:

https://www.anyscale.com/blog/fine-tuning-llama-2-a-comprehensive-case-study-for-tailoring-models-to-unique-applications

Fine Tuning LlaMa-2

Although LlaMa-2 is an improvement over its predecessor LlaMa, you can further improve the performance of Llama-2 by performing some fine tuning of this LLM.

https://medium.com/@murtuza753/using-llama-2-0-faiss-and-langchain-for-question-answering-on-your-own-data-682241488476

The following article shows you how to fine tune LlaMa-2 in a Google Colaboratory notebook:

https://towardsdatascience.com/fine-tune-your-own-llama-2-model-in-a-colab-notebook-df9823a04a32

The following article describes how to use MonsterAPI (also discussed in the article) in order to fine tune LlaMa-2 in five steps:

https://blog.monsterapi.ai/how-to-fine-tune-llama-2-llm/

The following link describes how to access LlaMa-2 in Google Colaboratory: *https://levelup.gitconnected.com/harnessing-the-power-of-llama-2-using-google-colab-2e1dedc2d1d8*

"HELLO WORLD" FROM GPT-4

Now that you have a basic understanding of ChatGPT and GPT-4, let's look at the "Hello World" code sample using ChatGPT. Listing 8.1 displays the contents of the code sample HelloWorld.java that was generated by ChatGPT.

LISTING 8.1: HelloWorld.java

```
public class HelloWorld {
    public static void main(String[] args) {
        System.out.println("Hello, World!");
    }
}
```

Listing 8.1 is accompanied by the following explanation of the code that was generated by ChatGPT.

```
Explanation of the Code:
    Class Definition: The public class HelloWorld line starts
the definition of a new public class called HelloWorld.
    Main Method: public static void main(String[] args)
declares the main method, which is the entry point for the
Java application.
    System.out.println(): Inside the main method, System.
out.println("Hello, World!"); is used to print the string
"Hello, World!" to the console.

How to Run the Code:
Save the code in a file named HelloWorld.java.
Open a terminal and navigate to the directory where
HelloWorld.java is saved.
Compile the code using the javac command: javac HelloWorld.
java. This will create a HelloWorld.class file.
Run the compiled class using the java command: java HelloWorld.

You should see "Hello, World!" printed to the terminal.
```

A more detailed explanation for ChatGPT-generated `Java` code samples is provided in Chapter 9, which contains numerous `Java`-based code samples that were generated by ChatGPT.

WHEN IS GPT-5 AVAILABLE?

As this book goes to print, there is no official information available regarding the status of GPT-5, which is to say that everything is speculative. In the early part of 2023 Sam Altman (CEO of OpenAI) remarked that there were "no official plans" for GPT-5.

However, during mid-2023 OpenAI filed a patent for GPT-5 in which there are some high-level details about the features of GPT-5. Some people have speculated that GPT-5 will be a more powerful version of GPT-4, and others suggest that filing a patent might be nothing more than securing the name GPT-5 by OpenAI.

Regardless of the motivation for filing a patent, there is a great deal of competition with GPT-4 from various companies. Therefore, it's highly plausible that OpenAI will release GPT-5, perhaps by the end of 2023. Regarding model sizes, recall that GPT-3 has 175B parameters, and some speculate that GPT-4 has 10 trillion parameters, which would mean that GPT-4 is roughly 60 times larger than GPT-3. The same increase in scale for GPT-5 seems implausible because GPT-5 would then consist of 600 trillion parameters.

Another possibility is that GPT-4 is based on the MoE (mixture of experts) methodology that involves multiple components. For instance, GPT-4 could be a combination of 8 components, each of which involves 220 million parameters, and therefore GPT-4 would consist of 1.76 trillion parameters.

Keep in mind that training LLMs such as GPT-4 is very costly and requires huge datasets for the pre-training step. Regardless of the eventual size of GPT-5, the training process could involve enormous costs.

SUMMARY

This chapter started with a discussion of ChatGPT from OpenAI and some of its features. In addition, you will learn about some competitors to ChatGPT, such as Claude-2 from Anthropic.

Next you learned about GPT-4 from OpenAI, which powers ChatGPT, and some of its features. Then you learned about some competitors of GPT-4, such as Llama-2 (Meta) and Bard (Google).

CHAPTER 9

JAVA AND GPT-4

This chapter contains an assortment of Java code samples that have been generated by ChatGPT using the "Code Interpreter" plugin in order to gain access to GPT-4. Keep in mind that the code samples in this chapter are displayed in a slightly modified manner from the actual output that you will see in ChatGPT. Specifically, some cosmetic changes have been made, such as adding section headings, in order to provide a smoother flow for the explanations that follow the code samples. In general, the layout of the code and the explanation for the code is very similar to what you will see in ChatGPT, which will prepare you for what to expect from ChatGPT.

Furthermore, the coding style might be different from the coding style that you prefer for your code. One immediate difference involves the placement of the "{" character on the same line as the class definition or the method definitions. However, the code samples provide an opportunity to see coding conventions that might be different from your coding conventions. Indeed, after finishing this chapter you can compare the code samples in this chapter with the corresponding hand-crafted counterparts in previous chapters.

The first part of this chapter contains ChatGPT-generated Java code samples for reversing a string with recursion, calculating factorial values, checking for prime numbers, and checking Goldbach's conjecture.

The second part of this chapter checks if a string is a palindrome, implements a binary search algorithm, and checks for unique characters in a string. The third part of this chapter inserts strings into a linked list and removes duplicate strings from a linked list. The final portion of this chapter shows you how to implement a queue using an array as well as how to implement a stack using an array.

One other detail pertains to the prompts that are provided to ChatGPT in order to generate Java code. For example, the format for the prompt pertaining to a binary search is "Please generate a Java class that implements a binary search." Other code samples are generated using similar prompts, where you would replace "binary search" with a different algorithm. Indeed, the prompts for ChatGPT are surprisingly simple in order to generate the Java code in this chapter.

REVERSE A STRING USING RECURSION

Listing 9.1 displays the contents of `ReverseString.java` that shows the code that ChatGPT generated in order to reverse a string via recursion.

LISTING 9.1: ReverseString.java

```
public class ReverseString {
    public static String reverseString(String str) {
        // Base case: if the string is empty or has only
        one character
        if (str.length() <= 1) {
            return str;
        }

        // Recursive case: reverse the substring
        // and append the first character at the end
        return reverseString(str.substring(1)) + str.
        charAt(0);
    }

    public static void main(String[] args) {
        String originalString = "hello";
        String reversedString = reverseString(originalString);

        System.out.println("Original String: " + originalString);
        System.out.println("Reversed String: " + reversedString);
    }
}
```

Listing 9.1 is accompanied by the following explanation of the code that was generated by ChatGPT.

```
Explanation of the Code:
Method Definition: The public static String
reverseString(String str) method is responsible for
reversing the input string str.

Base Case: The base case for the recursion is when the
string length is 1 or 0. In this case, the string itself is
returned as it is already "reversed."

if (str.length() <= 1) {
    return str;
}
```

```
Recursive Case: The recursive part takes the substring of
the string (excluding the first character) and reverses it.
It then appends the first character at the end.

return reverseString(str.substring(1)) + str.charAt(0);

Main Method: The main method is the entry point for the
Java program. It calls the reverseString function and
prints both the original and reversed strings.

How to Run the Code:
    Save the code in a file named ReverseString.java.
    Open a terminal and navigate to the directory where
ReverseString.java is saved.
    Compile the code using the javac command: javac
ReverseString.java.
    Run the compiled class using the java command: java
ReverseString.

You should see the original and reversed strings printed to
the terminal.
```

The output from Listing 9.1 is here:

```
Original String: hello

Reversed String: olleh
```

CALCULATE FACTORIAL VALUES

Listing 9.2 displays the contents of `Factorial.java` that shows the code that ChatGPT generated in order to calculate factorial values.

LISTING 9.2: Factorial.java

```
public class Factorial {

    // Iterative method to calculate factorial
    public static int factorialIterative(int n) {
        int result = 1;
        for (int i = 1; i <= n; i++) {
            result *= i;
        }
        return result;
    }

    // Recursive method to calculate factorial
    public static int factorialRecursive(int n) {
        if (n == 0) {
            return 1;
        }
```

```
        return n * factorialRecursive(n - 1);
    }

    public static void main(String[] args) {
        // Test the iterative method
        int n = 5;
        int resultIterative = factorialIterative(n);
        System.out.println("Factorial of " + n + "
        (iterative): " + resultIterative);

        // Test the recursive method
        int resultRecursive = factorialRecursive(n);
        System.out.println("Factorial of " + n + "
        (recursive): " + resultRecursive);
    }
}
```

Listing 9.2 is accompanied by the following explanation of the code that was generated by ChatGPT.

```
Explanation of the Code
    Import Statements: There are no import statements in
this example because we're using only Java's built-in
functionality.
    Class Definition: The class Factorial is defined, which
will contain all our methods.

public class Factorial {

Iterative Factorial Method: The factorialIterative(int
n) method calculates the factorial of an integer nn
iteratively.

public static int factorialIterative(int n) {
    int result = 1;
    for (int i = 1; i <= n; i++) {
        result *= i;
    }
    return result;
}

We initialize a variable result to 1.
We then loop from 1 to nn, multiplying result by each
integer.
The method returns result.

Recursive Factorial Method: The factorialRecursive(int
n) method calculates the factorial of an integer nn
recursively.

public static int factorialRecursive(int n) {
```

```
    if (n == 0) {
        return 1;
    }
    return n * factorialRecursive(n - 1);
}
```

```
If nn is 0, it returns 1 (base case).
Otherwise, it returns n×factorial of (n-1)
n×factorial of (n-1).

Main Method: The main(String[] args) method is the entry
point of the program.
    public static void main(String[] args) {
        int n = 5;
        int resultIterative = factorialIterative(n);
        System.out.println("Factorial of " + n + "
        (iterative): " + resultIterative);

        int resultRecursive = factorialRecursive(n);
        System.out.println("Factorial of " + n + "
        (recursive): " + resultRecursive);
    }

        It calls both the factorialIterative and
factorialRecursive methods and prints the results.

You can run this code in any Java environment. It
calculates the factorial of 5 both iteratively and
recursively, and then it prints the results.
```

The output from Listing 9.2 is here:

```
Factorial of 5 (iterative): 120

Factorial of 5 (recursive): 120
```

CHECK FOR PRIME NUMBERS

Listing 9.3 displays the contents of `PrimeCheck.java` that shows the code that ChatGPT generated in order to reverse a string via recursion.

LISTING 9.3: PrimeCheck.java

```
public class PrimeCheck {

    // Function to check if a number is prime
    public static boolean isPrime(int n) {
        // Base case: 0 and 1 are not prime numbers
        if (n <= 1) {
            return false;
        }
```

```
        // Check for factors other than 1 and n
        for (int i = 2; i <= Math.sqrt(n); i++) {
            if (n % i == 0) {
                // n has a divisor other than 1 and itself
                return false;
            }
        }

        return true; // n is a prime number
    }

    public static void main(String[] args) {
        // Test numbers
       int[] testNumbers = {1, 2, 3, 4, 5, 16, 17, 18, 19, 20};

        for (int num : testNumbers) {
            if (isPrime(num)) {
               System.out.println(num + " is a prime number.");
            } else {
                System.out.println(num + " is not a prime
                number.");
            }
        }
    }
}
```

Listing 9.3 is accompanied by the following explanation of the code that was generated by ChatGPT.

```
Explanation of the Code:
    Method Definition: The public static boolean isPrime(int n)
method is responsible for checking if the input integer n is a
prime number.

    Base Case: The base case checks if n is less than or
equal to 1. If so, n is not a prime number.

if (n <= 1) {
    return false;
}

Loop for Checking Divisors: We use a for-loop that runs
up to the square root of n. This optimization reduces the
number of iterations needed to check for divisors.

    for (int i = 2; i <= Math.sqrt(n); i++) {
        if (n % i == 0) {
            return false;
        }
    }
```

```
    Main Method: The main method serves as the entry point
for the program. It initializes an array of test numbers
and uses the isPrime function to check if each number is
prime, printing the result.

How to Run the Code:

    Save the code in a file named PrimeCheck.java.
    Open a terminal and navigate to the directory where
PrimeCheck.java is saved.
    Compile the code using the javac command: javac
PrimeCheck.java.
    Run the compiled class using the java command: java
PrimeCheck.

You should see messages printed to the terminal indicating
whether each test number is a prime number or not.
```

The output from Listing 9.3 is here:

```
1 is not a prime number.
2 is a prime number.
3 is a prime number.
4 is not a prime number.
5 is a prime number.
16 is not a prime number.
17 is a prime number.
18 is not a prime number.
19 is a prime number.
20 is not a prime number.
```

CHECKING GOLDBACH'S CONJECTURE

Listing 9.4 displays the contents of `GoldbachConjecture.java` that shows the code that ChatGPT generated in order to reverse a string via recursion.

LISTING 9.4: GoldbachConjecture.java

```
// Goldbach's Conjecture states that every even integer
greater than 2 can be expressed as the sum of two prime
numbers. Below is a Java code snippet that demonstrates how
to check Goldbach's Conjecture for a given even integer nn.

public class GoldbachConjecture {

    // Function to check if a number is prime
    public static boolean isPrime(int n) {
        if (n <= 1) {
            return false;
        }
```

```
        for (int i = 2; i <= Math.sqrt(n); i++) {
            if (n % i == 0) {
                return false;
            }
        }

        return true;
    }

    // Check Goldbach's Conjecture for a given even number n
    public static void checkGoldbach(int n) {
        if (n <= 2 || n % 2 != 0) {
            System.out.println("Please enter an even number
            greater than 2.");
            return;
        }

        for (int i = 2; i <= n / 2; i++) {
            int j = n - i;

            if (isPrime(i) && isPrime(j)) {
               System.out.println(n + " = " + i + " + " + j);
                return;
            }
        }

        System.out.println("Goldbach's Conjecture is
        incorrect for " + n);
    }

    public static void main(String[] args) {
        // Test Goldbach's Conjecture for even numbers from
        4 to 20
        for (int i = 4; i <= 20; i += 2) {
            checkGoldbach(i);
        }
    }
}
```

Listing 9.4 is accompanied by the following explanation of the code that was generated by ChatGPT.

```
Explanation of the Code:
    isPrime Method: This method checks if a given number
nn is prime or not, using the same logic from the previous
example.

    checkGoldbach Method: This method checks Goldbach's
Conjecture for a given even integer nn.
        First, it checks if nn is even and greater than 2.
If not, it prints an error message.
```

```
        It then iterates through all integers ii from 2 to
n22n<200b>, checking if both ii and n-in-i are prime. If
such a pair is found, it prints the pair and returns.

    Main Method: The main method serves as the entry point
for the program. It tests Goldbach's Conjecture for even
numbers from 4 to 20.

How to Run the Code:

    Save the code in a file named GoldbachConjecture.java.
    Open a terminal and navigate to the directory where
GoldbachConjecture.java is saved.
    Compile the code using the javac command: javac
GoldbachConjecture.java.
    Run the compiled class using the java command: java
GoldbachConjecture.

You should see messages printed to the terminal showing how
each even number from 4 to 20 can be expressed as the sum
of two prime numbers, in line with Goldbach's Conjecture.
```

The output from Listing 9.4 is here:

```
1 is not a prime number.
2 is a prime number.
3 is a prime number.
4 is not a prime number.
5 is a prime number.
16 is not a prime number.
17 is a prime number.
18 is not a prime number.
19 is a prime number.
20 is not a prime number.
```

CHECK FOR PALINDROMES

Listing 9.5 displays the contents of `PalindromeChecker.java` that shows the code that ChatGPT generated in order to check if a string is a palindrome.

LISTING 9.5: PalindromeChecker.java

```
    public static boolean isPalindrome(String str) {
        // Remove case sensitivity and spaces for a more
        robust check
        str = str.toLowerCase().replaceAll("\\s+", "");

        // Base case: a string with 0 or 1 characters is a
        palindrome
        if (str.length() <= 1) {
            return true;
```

```
        }

        // Compare the first and last characters
        if (str.charAt(0) != str.charAt(str.length() - 1))
{
            return false;
        }

        // Recursive case: Check the substring that
        // excludes the first and last characters
       return isPalindrome(str.substring(1, str.length() - 1));
    }

    public static void main(String[] args) {
        String testString1 = "radar";
        String testString2 = "hello";
        String testString3 = "A man a plan a canal Panama";

        System.out.println(testString1 + " is a palindrome:
        " + isPalindrome(testString1));
        System.out.println(testString2 + " is a palindrome:
        " + isPalindrome(testString2));
        System.out.println(testString3 + " is a palindrome:
        " + isPalindrome(testString3));
    }
}
```

Listing 9.5 is accompanied by the explanation of the code (shown in the next section) that was generated by ChatGPT, and the output from Listing 9.5 is here:

```
radar is a palindrome: true

hello is a palindrome: false
```

IMPLEMENT A BINARY SEARCH ALGORITHM

Listing 9.6 displays the contents of `BinarySearch.java` that shows the code that ChatGPT generated in order to perform a binary search.

LISTING 9.6: BinarySearch.java

```
public class BinarySearch {

    public static int binarySearch(int[] arr, int target,
    int left, int right) {
        if (left > right) {
            return -1; // Target not found
        }
```

```
        // Calculate mid index: see comment after the code
        int mid = left + (right - left) / 2;

        if (arr[mid] == target) {
            // Target found at index mid
            return mid;
        } else if (arr[mid] < target) {
            // Search the right half
            return binarySearch(arr, target, mid + 1,
            right);
        } else {
            // Search the left half
            return binarySearch(arr, target, left, mid - 1);
        }
    }

    public static void main(String[] args) {
        int[] arr = {1, 3, 5, 7, 9, 11, 13, 15, 17, 19};
        int target = 15;

        int result = binarySearch(arr, target, 0, arr.
        length - 1);

        if (result == -1) {
            System.out.println("Target " + target + " not
            found in the array.");
        } else {
            System.out.println("Target " + target + " found
            at index " + result + ".");
        }
    }
}
```

Listing 9.6 is accompanied by the following explanation of the code that was generated by ChatGPT.

```
Explanation of the Code:
    Method Definition: The method public static int
binarySearch(int[] arr, int target, int left, int right)
is responsible for performing the binary search. It takes
an integer array arr, the target integer to find, and two
indices left and right that define the range within which to
search.

Base Case: If left is greater than right, the target was
not found, and the method returns -1.

if (left > right) {
    return -1;
}
```

```
Calculate Midpoint: The midpoint index mid is calculated
using the formula mid=left+right-left2mid=left+2right-left.
int mid = left + (right - left) / 2;

Comparison and Recursion: Depending on whether the element
at the midpoint is less than, equal to, or greater than the
target, the function either returns the index or continues
to search in the left or right half of the array.

    if (arr[mid] == target) {
        return mid;
    } else if (arr[mid] < target) {
        return binarySearch(arr, target, mid + 1, right);
    } else {
        return binarySearch(arr, target, left, mid - 1);
    }

Main Method: The main method serves as the entry point for
the program. It initializes an array and a target, and
then calls the binarySearch method. Finally, it prints the
result.

How to Run the Code:
    Save the code in a file named BinarySearch.java.
    Open a terminal and navigate to the directory where
BinarySearch.java is saved.
    Compile the code using the javac command: javac
BinarySearch.java.
    Run the compiled class using the java command: java
BinarySearch.

You should see the result of the binary search printed to
the terminal.
```

The output from Listing 9.6 is here:

```
Target 15 found at index 7.
```

CHECK FOR UNIQUE CHARACTERS IN A STRING

Listing 9.7 displays the contents of `ReverseString.java` that shows the code that ChatGPT generated in order to determine whether or not a string contains unique characters.

LISTING 9.7: UniqueCharacters1.java

```
// Below is a Java code snippet that demonstrates how to
check if a given string contains unique characters. The
example includes two approaches: one that uses a HashSet
```

```
for storage and one that does not use any additional data
structures.

// Using HashSet to Check for Unique Characters
public class UniqueCharacters1 {

    public static boolean hasUniqueCharacters(String str) {
        HashSet<Character> charSet = new HashSet<>();

        for (char c : str.toCharArray()) {
            if (charSet.contains(c)) {
                return false;
            }
            charSet.add(c);
        }

        return true;
    }

    public static void main(String[] args) {
        String[] testStrings = {"apple", "banana",
        "cherry", "dog", "unique"};

        for (String s : testStrings) {
            System.out.println("\"" + s + "\" has unique
            characters: " + hasUniqueCharacters(s));
        }
    }
}
```

Listing 9.7 is accompanied by the following explanation of the code that was generated by ChatGPT.

```
Explanation of the Code:

    Method Definition: The public static boolean
hasUniqueCharacters(String str) method checks if the input
string str has unique characters.

    Using HashSet: The first approach uses a HashSet to
store characters as we iterate through the string. If
we find a character already in the set, we return false.
Otherwise, we add the character to the set.

    Without Additional Data Structures: The second approach
uses two nested loops to compare each character in the
string with every other character. This method is less
efficient (O(n2)O(n2)) but does not use any additional data
structures.
```

```
    Main Method: The main method serves as the entry point
for the program. It tests a few example strings and prints
whether each string has unique characters.

How to Run the Code:

    Save one of the code versions in a file named
UniqueCharacters.java.
    Open a terminal and navigate to the directory where
UniqueCharacters.java is saved.
    Compile the code using the javac command: javac
UniqueCharacters.java.
    Run the compiled class using the java command: java
UniqueCharacters.

You should see messages printed to the terminal indicating
whether each test string has unique characters or not.
```

The output from Listing 9.7 is here:

```
"apple" has unique characters: false
"banana" has unique characters: false
"cherry" has unique characters: false
"dog" has unique characters: true
"unique" has unique characters: false
```

Listing 9.8 displays the contents of `UniqueCharacters2.java` that shows the code that ChatGPT generated in order to determine whether or not a string contains unique characters.

LISTING 9.8: UniqueCharacters2.java

```
// Below is a Java code snippet that demonstrates how to
check if a given string contains unique characters. The
example includes two approaches: one that uses a HashSet
for storage and one that does not use any additional data
structures.

//Using HashSet to Check for Unique Characters
import java.util.HashSet;

// Without Using Additional Data Structures
public class UniqueCharacters2 {

    public static boolean hasUniqueCharacters(String str) {
        for (int i = 0; i < str.length(); i++) {
            for (int j = i + 1; j < str.length(); j++) {
                if (str.charAt(i) == str.charAt(j)) {
                    return false;
                }
            }
```

```
        }

        return true;
    }

    public static void main(String[] args) {
        String[] testStrings = {"apple", "banana",
        "cherry", "dog", "unique"};

        for (String s : testStrings) {
            System.out.println("\"" + s + "\" has unique
            characters: " + hasUniqueCharacters(s));
        }
    }
}
```

Listing 9.8 is accompanied by the following explanation of the code that was generated by ChatGPT.

```
Explanation of the Code:

    Method Definition: The public static boolean
hasUniqueCharacters(String str) method checks if the input
string str has unique characters.

    Using HashSet: The first approach uses a HashSet to
store characters as we iterate through the string. If
we find a character already in the set, we return false.
Otherwise, we add the character to the set.

    Without Additional Data Structures: The second approach
uses two nested loops to compare each character in the
string with every other character. This method is less
efficient (O(n2)O(n2)) but does not use any additional data
structures.

    Main Method: The main method serves as the entry point
for the program. It tests a few example strings and prints
whether each string has unique characters.

How to Run the Code:

    Save one of the code versions in a file named
UniqueCharacters.java.
    Open a terminal and navigate to the directory where
UniqueCharacters.java is saved.
    Compile the code using the javac command: javac
UniqueCharacters.java.
    Run the compiled class using the java command: java
UniqueCharacters.

You should see messages printed to the terminal indicating
whether each test string has unique characters or not.
```

The output from Listing 9.8 is here:

```
"apple" has unique characters: false
"banana" has unique characters: false
"cherry" has unique characters: false
"dog" has unique characters: true
"unique" has unique characters: false
```

INSERT STRINGS IN A LINKED LIST

Listing 9.9 displays the contents of `StringLinkedListExample.java` that shows the code that ChatGPT generated in order to insert strings in a linked list.

LISTING 9.9: StringLinkedListExample

```
public class StringLinkedListExample {

/*
Below is a simple Java code snippet that demonstrates
inserting strings into a linked list using Java's built-in
LinkedList class from the java.util package. Additionally,
I've implemented a custom linked list for string insertion.
Using Java's Built-in LinkedList
*/

import java.util.LinkedList;

public class StringLinkedListExample {
    public static void main(String[] args) {
        // Create a new LinkedList of strings
        LinkedList<String> stringList = new LinkedList<>();

        // Insert strings into the LinkedList
        stringList.add("Apple");
        stringList.add("Banana");
        stringList.add("Cherry");

        // Display the LinkedList
        System.out.println("Built-in LinkedList: " +
        stringList);
    }
}

// Using Custom Linked List
public class CustomStringLinkedList {
    Node head;

    static class Node {
        String data;
        Node next;
```

```
        Node(String data) {
            this.data = data;
            next = null;
        }
    }

    // Function to insert a new node
    public static CustomStringLinkedList
    insert(CustomStringLinkedList list, String data) {
        Node newNode = new Node(data);
        newNode.next = null;

        if (list.head == null) {
            list.head = newNode;
        } else {
            Node last = list.head;
            while (last.next != null) {
                last = last.next;
            }
            last.next = newNode;
        }
        return list;
    }

    // Function to print the LinkedList
    public static void printList(CustomStringLinkedList list) {
        Node currNode = list.head;

        System.out.print(„Custom LinkedList: „);

        while (currNode != null) {
            System.out.print(currNode.data + " -> ");
            currNode = currNode.next;
        }
        System.out.println(„null");
    }

    public static void main(String[] args) {
        CustomStringLinkedList list = new
        CustomStringLinkedList();

        // Insert strings into custom LinkedList
        insert(list, "Apple");
        insert(list, "Banana");
        insert(list, "Cherry");

        // Display custom LinkedList
        printList(list);
    }
}
```

Listing 9.9 is accompanied by the following explanation of the code that was generated by ChatGPT.

```
Explanation of the Code
Built-in LinkedList: The first example uses Java's built-in
LinkedList class to add strings. The add() method appends
the elements to the end of the list.

Custom Linked List: The second example demonstrates a
simple custom implementation of a linked list that stores
strings.
        Node: A nested static class representing a node in
the linked list. Each Node has a data field for storing the
string and a next field for referencing the next node in the
list.
        insert(): A method that takes a string and adds a
new node with that string to the end of the list.
        printList(): A method that prints the entire list.

How to Run the Code
    Save the code in a file (either as two separate
files like StringLinkedListExample.java and
CustomStringLinkedList.java or combined in one file).
    Open a terminal and navigate to the directory where the .java
file(s) are saved.
    Compile the code using the javac command:
javac StringLinkedListExample.java or javac
CustomStringLinkedList.java.
    Run the compiled class using the java command: java
StringLinkedListExample or java CustomStringLinkedList.

You should see the linked lists printed to the terminal.
```

The output from Listing 9.9 is here:

```
Built-in LinkedList: [Apple, Banana, Cherry]
```

REMOVE DUPLICATE STRINGS FROM A LINKED LIST

Listing 9.10 displays the contents of `RemoveDuplicatesLinkedList.java` that shows the code that ChatGPT generated in order to remove duplicate strings from a linked list.

LISTING 9.10: RemoveDuplicatesLinkedList.java

```
public class Factorial {

import java.util.LinkedList;
import java.util.HashSet;

public class RemoveDuplicatesLinkedList {
```

```
    public static void main(String[] args) {
        // Create a LinkedList with duplicates
        LinkedList<Integer> list = new LinkedList<>();
        list.add(1);
        list.add(2);
        list.add(3);
        list.add(3);
        list.add(4);
        list.add(4);
        list.add(5);

        System.out.println("Original LinkedList: " + list);

        // Remove duplicates
        removeDuplicates(list);

        System.out.println("LinkedList after removing
        duplicates: " + list);
    }

    public static void removeDuplicates(LinkedList<Integer>
    list) {
        HashSet<Integer> seen = new HashSet<>();
        for (int i = 0; i < list.size(); i++) {
            int value = list.get(i);
            if (seen.contains(value)) {
                list.remove(i);
               i--; // Decrement i as the list size has changed
            } else {
                seen.add(value);
            }
        }
    }
}
```

Listing 9.10 is accompanied by the following explanation of the code that was generated by ChatGPT.

```
Explanation of the Code
Initialization: We initialize a LinkedList named list and
populate it with some integer values, including duplicates.
LinkedList<Integer> list = new LinkedList<>();
list.add(1);
list.add(2);
list.add(3);
list.add(3);
list.add(4);
list.add(4);
list.add(5);
```

```
Remove Duplicates: The removeDuplicates method removes
duplicate elements from the linked list. It uses a HashSet
to keep track of elements that have already been seen.
    public static void removeDuplicates(LinkedList<Integer>
    list) {
        HashSet<Integer> seen = new HashSet<>();
        for (int i = 0; i < list.size(); i++) {
            int value = list.get(i);
            if (seen.contains(value)) {
                list.remove(i);
                i--; // Decrement i as the list size has changed
            } else {
                seen.add(value);
            }
        }
    }

        We use a HashSet named seen to store unique
elements as we traverse the linked list.
        We iterate through the LinkedList and check each
element.
        If an element is already in seen, we remove it from
the LinkedList.
        Otherwise, we add the element to seen.

Main Method: The main method serves as the entry point
for the program. It initializes a LinkedList, calls the
removeDuplicates method to remove duplicates, and then
prints the updated list.

How to Run the Code
    Save the code in a file named
RemoveDuplicatesLinkedList.java.
    Open a terminal and navigate to the directory where
RemoveDuplicatesLinkedList.java is saved.
    Compile the code using the javac command: javac
RemoveDuplicatesLinkedList.java.
    Run the compiled class using the java command: java
RemoveDuplicatesLinkedList.
You should see the original and updated linked lists
printed to the terminal, showing that duplicates have been
removed.
```

The output from Listing 9.10 is here:

```
Original LinkedList: [1, 2, 3, 3, 4, 4, 5]
LinkedList after removing duplicates: [1, 2, 3, 4, 5]
```

IMPLEMENT A QUEUE USING AN ARRAY

Listing 9.11 displays the contents of `ArrayQueue.java` that shows the code that ChatGPT generated in order to implement a queue using an array.

LISTING 9.11: ArrayQueue.java

```
// Below is a Java code snippet that demonstrates how to
implement a simple queue using an array. This example
includes methods to insert values (enqueue) and remove
values (dequeue) from the queue.

public class ArrayQueue {
    private int capacity;
    private int size;
    private int front;
    private int rear;
    private int[] array;

    public ArrayQueue(int capacity) {
        this.capacity = capacity;
        this.size = 0;
        this.front = 0;
        this.rear = capacity - 1;
        this.array = new int[this.capacity];
    }

    // Check if the queue is full
    public boolean isFull() {
        return size == capacity;
    }

    // Check if the queue is empty
    public boolean isEmpty() {
        return size == 0;
    }

    // Add an element to the queue
    public void enqueue(int item) {
        if (isFull()) {
            System.out.println("Queue is full. Cannot
            enqueue.");
            return;
        }
        rear = (rear + 1) % capacity;
        array[rear] = item;
        size++;
        System.out.println("Enqueued: " + item);
    }

    // Remove an element from the queue
    public int dequeue() {
        if (isEmpty()) {
            System.out.println("Queue is empty. Cannot
            dequeue.");
            return Integer.MIN_VALUE;
```

```
        }
        int item = array[front];
        front = (front + 1) % capacity;
        size--;
        return item;
    }

    public static void main(String[] args) {
        ArrayQueue queue = new ArrayQueue(5);

        // Enqueue elements
        queue.enqueue(1);
        queue.enqueue(2);
        queue.enqueue(3);
        queue.enqueue(4);
        queue.enqueue(5);

        // Try to enqueue when queue is full
        queue.enqueue(6);

        // Dequeue elements
        System.out.println(„Dequeued: „ + queue.dequeue());
        System.out.println(„Dequeued: „ + queue.dequeue());

        // Enqueue more elements
        queue.enqueue(6);
        queue.enqueue(7);
    }
}
```

Listing 9.11 is accompanied by the following explanation of the code that was generated by ChatGPT.

```
Explanation of the Code
    Initialization: The ArrayQueue class is initialized
with a capacity, size, front, rear, and an array named
array to hold the queue elements.

    isFull and isEmpty: These methods check whether the
queue is full or empty, respectively.

    enqueue: The enqueue method adds an element to the rear
of the queue. If the queue is full, it prints a message and
returns without adding the element.

    dequeue: The dequeue method removes an element from the
front of the queue and returns it. If the queue is empty,
it prints a message and returns Integer.MIN_VALUE.

    Main Method: The main method serves as the entry point
for the program. It creates an ArrayQueue object with a
```

```
capacity of 5, enqueues some elements, and then dequeues
some.

How to Run the Code
    Save the code in a file named ArrayQueue.java.
    Open a terminal and navigate to the directory where
ArrayQueue.java is saved.
    Compile the code using the javac command: javac
ArrayQueue.java.
    Run the compiled class using the java command: java
ArrayQueue.

You should see messages printed to the terminal indicating
the elements being enqueued and dequeued, as well as any
errors if the queue is full or empty.
```

The output from Listing 9.11 is here:

```
Enqueued: 1
Enqueued: 2
Enqueued: 3
Enqueued: 4
Enqueued: 5
Queue is full. Cannot enqueue.
Dequeued: 1
Dequeued: 2
Enqueued: 6
Enqueued: 7
```

IMPLEMENT A STACK USING AN ARRAY

Listing 9.12 displays the contents of `ArrayStack.java` that shows the code that ChatGPT generated in order to implement a stack using an array.

LISTING 9.x: ArrayStack.java

```
// Below is a Java code snippet that demonstrates how to
implement a stack using an array. The example includes
methods to insert values (push) and remove values (pop)
from the stack.

public class ArrayStack {
    private int capacity;
    private int top;
    private int[] array;

    public ArrayStack(int capacity) {
        this.capacity = capacity;
        this.top = -1;
        this.array = new int[capacity];
    }
```

```
    // Check if the stack is full
    public boolean isFull() {
        return top == capacity - 1;
    }

    // Check if the stack is empty
    public boolean isEmpty() {
        return top == -1;
    }

    // Add an element to the stack
    public void push(int item) {
        if (isFull()) {
            System.out.println("Stack is full. Cannot push.");
            return;
        }
        array[++top] = item;
        System.out.println("Pushed: " + item);
    }

    // Remove an element from the stack
    public int pop() {
        if (isEmpty()) {
            System.out.println("Stack is empty. Cannot pop.");
            return Integer.MIN_VALUE;
        }
        return array[top--];
    }

    public static void main(String[] args) {
        ArrayStack stack = new ArrayStack(5);

        // Push elements onto the stack
        stack.push(1);
        stack.push(2);
        stack.push(3);
        stack.push(4);
        stack.push(5);

        // Try to push when stack is full
        stack.push(6);

        // Pop elements from the stack
        System.out.println("Popped: " + stack.pop());
        System.out.println("Popped: " + stack.pop());

        // Push more elements
        stack.push(6);
        stack.push(7);
    }
}
```

Listing 9.12 is accompanied by the following explanation of the code that was generated by ChatGPT.

```
Explanation of the Code
    Initialization: The ArrayStack class is initialized
with a capacity, a top index initialized to -1, and an
array named array to hold the stack elements.

    isFull and isEmpty: These methods check whether the
stack is full or empty, respectively.

    push: The push method adds an element to the top of
the stack. If the stack is full, it prints a message and
returns without adding the element.

    pop: The pop method removes an element from the top of
the stack and returns it. If the stack is empty, it prints
a message and returns Integer.MIN_VALUE.

    Main Method: The main method serves as the entry point
for the program. It creates an ArrayStack object with a
capacity of 5, pushes some elements onto it, and then pops
some.

How to Run the Code

    Save the code in a file named ArrayStack.java.
    Open a terminal and navigate to the directory where
ArrayStack.java is saved.
    Compile the code using the javac command: javac
ArrayStack.java.
    Run the compiled class using the java command: java
ArrayStack.

You should see messages printed to the terminal indicating
the elements being pushed and popped, as well as any errors
if the stack is full or empty.
```

The output from Listing 9.12 is here:

```
Pushed: 1
Pushed: 2
Pushed: 3
Pushed: 4
Pushed: 5
Stack is full. Cannot push.
Popped: 5
Popped: 4
Pushed: 6
Pushed: 7
```

SUMMARY

This chapter introduced you to Java code samples that were generated by ChatGPT using the "Code Interpreter" plugin.

The first example involves reversing a string with recursion, followed by an example of calculating factorial values. Then you learned how to check for prime numbers and how to check the validity of Goldbach's conjecture.

In addition, you saw Java code that checks if a string is a palindrome, as well as how to implement a binary search algorithm. Next, you learned how to check for unique characters in a string.

Furthermore, you saw how to insert strings into a linked list and then how to remove duplicate strings from a linked list. Finally, you learned how to implement a queue using an array as well as how to implement a stack using an array.

INDEX

www.ingramcontent.com/pod-product-compliance
Lightning Source LLC
Chambersburg PA
CBHW060807220326
41598CB00022B/2557

* 9 7 8 1 5 0 1 5 2 2 4 3 7 *